Protein-Calorie Malnutrition

Protein-Calorie Malnutrition

A Nestlé Foundation Symposium

Editor

A. von Muralt

In collaboration with

H. Aebi, Berne; G. Arroyave, Guatemala; J. M. Bengoa, Geneva;
D. Bovet, Sassari; J. S. Dinning, Bangkok; G. Fanconi, Zurich;
S. Frenk, Mexico; C. Gopalan, Hyderabad; A. E. Harper, Madison;
H. Isliker, Lausanne; D. B. Jelliffe, Kingston; C. G. King, New York;
M. L. L. Martineaud, Yaoundé; J. Mauron, Vevey; E. M. Mrak, Davis;
Vera Mrak, Davis; L. R. Rey, Vevey; R. G. Whitehead, Kampala;
N. C. Wright, London; Chr. Zbinden, Vevey

Editorial staff S. Herzen, M. Frochaux, Cl. Meylan, N. Mercier,
M. Mingard

With 43 Figures

Springer-Verlag Berlin · Heidelberg · New York 1969

ISBN-13: 978-3-642-87971-5   e-ISBN-13: 978-3-642-87969-2
DOI: 10.1007/ 978-3-642-87969-2

List of Participants at the Symposium

Aebi, Dr. H., Professor of Biochemistry, Biochemical Institute, University of Berne, Bühlstraße 28, 3000 Berne (Switzerland).

Arroyave, Dr. G., Chief, Division of Physiological Chemistry, Instituto de Nutrición de Centro América y Panamá (I.N.C.A.P.), Apartado postal 1188, Guatemala-City (Guatemala).

Bengoa, Dr. J. M., Chief, Nutrition Unit, World Health Organisation, 1211 Geneva 27 (Switzerland).

Bovet, Dr. D., Nobel Prize, Professor of Pharmacology, Pharmacology Institute, University of Sassari, 07100 Sassari (Italy).

Dinning, Dr. J. S., The Rockefeller Foundation, G.P.O. 2453, Bangkok (Thailand).

Fanconi, Dr. G., Professor of Pediatrics emeritus, University of Zurich, Spiegelhofstrasse 39, 8032 Zurich (Switzerland).

Frenk, Dr. S., Nutrition Department, Instituto Mexicano de Seguro Social, Centro Médico Nacional, Hospital de Pediatría, México 7, D.F. (Mexico).

Gopalan, Dr. C., Director of the Nutrition Research Laboratories, President of the Committee on Procedures for Appraisal of Protein-Calorie Malnutrition, Tarnaka, Hyderabad-7, A.P. (India).

Harper, Dr. A. E., Professor of Biochemistry, University of Wisconsin, Madison, Wisconsin 53706 (U.S.A.).

Isliker, Dr. H., Professor of Biochemistry, Institute of Biochemistry, University of Lausanne, rue du Bugnon 21, 1000 Lausanne (Switzerland).

Jelliffe, Dr. D. B., Director, Caribbean Food and Nutrition Institute, P.O. Box 140, Kingston 7 (Jamaica, W.I.).

King, Dr. C. G., Professor and President of the International Union of Nutritional Sciences, Institute of Nutrition Sciences, Columbia University, 562 West 168th Street, New York 10032 (U.S.A.).

Martineaud, Dr. M. L. L., Nutritionist at W.H.O., Office of the Representative, P. O. B. 155, Yaoundé (Fed. Rep. of Cameroon)

Mauron, Dr. J., Department of Research and Development, Nestlé, 1800 Vevey (Switzerland).

Mrak, Dr. E. M., Professor of Food Technology, Chancellor, University of California, Davis, California 95616 (U.S.A.).

Mrak, Dr. Vera, Consultant to the Yolo Country Public Health Service, Davis, California 95616 (U.S.A.).

von Muralt, Dr. A., Professor of Physiology, President of Nestlé Foundation, Institute of Physiology, University of Berne, Bühlplatz 5, 3000 Berne (Switzerland).

Rey, Dr. L. R., Professor of Biochemistry, Nestlé Scientific Adviser, 1800 Vevey (Switzerland).

Whitehead, Dr. R. G., University of Cambridge, Cambridge (England) and Child Nutrition Research Unit, Box 7071 (MRC) Kampala (Uganda).

Wright, Sir Norman C., C. B., Secretary, British Association for the Advancement of Science, 65 Addison Road, London, W.14 (England).

Zbinden, Dr. Chr., Manager, Dietetic Products Department, Nestlé, 1800 Vevey (Switzerland).

Preface

Hunger is the world problem Nr 1, overshadowed by an uncontrollable explosion of the human population all over our planet. Lack of food has been one of the most primitive dangers, which animal life had to face at every stage of its evolution. The living body developed in the course of this evolution special emergency reactions against this danger, which is characterised by a lack of food calories, a lack of nitrogen in the form of proteins, a lack of vitamins and oligo elements. Based on an intricate physiological defense pattern man can support complete starvation up to one month, by using up the substance of less important organs in order to maintain the functional matrix of the important organs, mainly the brain and the nervous system. This regulation and the pattern of its mechanisms are of great interest to the physiologist who is aware that they are also responsible for the maintenance of life among millions of human beings who desperately live in a state of permanent hunger. The most serious problem in many developing countries is not the supply of calories (mainly carbohydrate calories) or the supply of vitamins and oligo elements, but the supply of a sufficient amount of protein in order to overcome the protein-calorie malnutrition. This problem must be considered as the most urgent one among all the other problems in the fight against hunger.

The aim of the Nestlé Foundation for the study of the problems of nutrition in the world is to further the improvement of nutrition by encouraging scientific and practical studies of problems directly connected with nutrition, in the areas of agricultural and animal production, food chemistry and technology, biology and physiology.

This is the reason why a first symposium on protein-calorie malnutrition was held in Lausanne, at the office of the Foundation, from September 6 to 7, 1968. It was very rewarding that a great number of distinguished experts in the field of protein-calorie malnutrition accepted the invitation of the Foundation and that the Committee on Procedures for Appraisal of Protein-Calorie Malnutrition of the International Union of Nutritional Sciences decided to hold its first meeting on the day preceding the symposium at the same place.

Biochemistry has advanced in the last decades in such a way that a great number of very precise micromethods are now available for the appraisal of protein-calorie malnutrition. The symposium was very encouraging for the plan that "biochemical batteries" should be used, not only in clinical work, but also in the studies on otherwise healthy population suffering from protein-calorie malnutrition. It was felt that a worldwide collaboration in this field, by using comparable methods, might help to appraise the deficits in protein and to measure the nutritional value of protein-rich new foodstuffs, not only in animal experiments but also with regard to their beneficial effects on malnourished human populations.

It is our hope that the publication of the papers and discussions of this first Nestlé Foundation Symposium will encourage many workers in the field of protein-calorie malnutrition to increase their efforts in fighting this deficiency, which seems to have also deleterious effects on the mental development.

With pleasure we record our thanks to all those who have contributed so actively to this symposium: the presidents of the meetings, the experts and guests, the members of the Council of our Foundation, the staff of our Foundation and last but not least the publisher of this book. We hope that this first book of a series of Nestlé Foundation publications will help to stimulate the interest of many workers in the field of protein-calorie malnutrition and thus help to fight this obstacle in the development of many countries. At this occasion we are also happy to extend our thanks to the initiators of our Foundation, Max Petitpierre, Enrico Bignami and Jean Constant Corthésy, who in a generous and far-sighted manner laid the basis for our work at the celebration of the centenary of Nestlé Alimentana.

A. von Muralt

Contents

From the Physiological Institute (Hallerianum) University of Berne (Switzerland)

Protein-Calorie Malnutrition Viewed as a Challenge for Homeostasis

A. von Muralt

Contents

Homeostasis

Protein-calorie malnutrition is a serious danger for the composition and size of the protein matrix in the human body, evoking physiological counter-reactions, which in turn tend to preserve this matrix, not in substance but with regard to its function. Such reactions are called "homeostatic reactions", leading to a functional adaptation under conditions of external stress. From the point of view of a physiologist, it is worthwhile to go back more than 40 years and to reconsider 6 postulates which were outlined by Walter Bradford Cannon in 1926, in a paper which unfortunately was buried in a Jubilee Volume in honour of Charles Richet (Cannon, 1926). I would like to quote these postulates in Cannon's own words:

"1. In an open system, such as our bodies represent, composed of unstable structure and subjected continually to disturbance, constancy is in itself evidence that agencies are acting or are ready to act to maintain this constancy.

2. If a homeostatic condition continues, it does so because any tendency towards a change is automatically met by increased effectiveness of a factor or factors which lessen the change.

3. A homeostatic agent does not act in opposite directions at the same point.

4. Homeostatic agents, antagonistic in one region of the body, may be cooperative in another region.

5. The regulating system which determines a homeostatic state may comprise a number of cooperating factors brought into action at the same time or successively.

6. When a factor is known which can shift a homeostatic state in one direction, it is reasonable to look for automatic control of that factor or for a factor or factors having an opposing effect."

At first sight one might be surprised to see these old postulates quoted in a discussion of protein-calorie malnutrition. The question arises, whether they have any bearing on this special problem. I would like to try to outline some considerations, which in my opinion justify viewing protein-calorie malnutrition as a challenge to homeostasis.

The Maintenance of Nitrogenous Equilibrium

Our body is able to vary the nitrogen excretion by a factor of 15 (30 — 2 g per day). The lower limit of regulation is necessary for the maintenance of the protein matrix, which consists of structural proteins and protein-stores (mainly in the liver). May I recall the classical data of Folin (1905) on the differences of composition of the urine on a diet rich or poor in protein, but of adequate caloric value. The volume of urine was reduced to $1/_3$, total nitrogen to $1/_5$, urea nitrogen to $1/_7$ (but in percentage of total nitrogen only a drop from 86% to 62%), uric acid nitrogen to $1/_2$, creatinine nitrogen remained *unchanged*, ammonia nitrogen remained *unchanged*. It is obvious that one or more homeostatic factors must have intervened to achieve this result: on the level of the kidney, a change in the pattern of tubular reabsorption and in the body, a change of protein breakdown, leading to a different pattern of concentration of the non-protein N-components of the blood plasma. The primitive concept of a simple relation between ingested and metabolised protein + metabolised tissue protein = excreted nitrogen × 6.25 is inadequate!

As a rule one assumes that excreted urea and half the uric acid are an index of exogenous metabolism or nitrogenous food intake; creatinine and half the uric acid should be an index of endogenous metabolism or tissue breakdown. But during starvation a considerable amount of urea is also excreted, which under these conditions is endogenous. These concepts of exogenous-N and endogenous-N are ambiguous and should be abolished! Urea nitrogen represents mainly the end product of protein, broken down for the purpose of *energy supply*. In starvation the missing food protein is replaced first by stored liver protein and then as a substitute by tissue protein, a homeostatic regulation directed towards the maintenance of an essential source of energy supply. But here the first question arises: where is the center for this regulation, how are these shifts to the mobilisation of liver protein and then to the selective breakdown of tissue proteins performed? Just to draw your attention closer to this point, I would like to recall that during starvation the tissues are sacrificed in inverse proportion to their "importance". The brain and heart lose only 3% of their bulk, the muscle 31%, the liver 54% and the spleen 67%. From the work of Hermann Rein (see H. Rein, 1948) and his school, we have learned that the oxygen supply to tissues is governed by their demand, but that in muscular exercise e.g. a great number of regulating mechanisms, hydrostatic, nervous and hormonal, are brought into play to ensure the oxygen supply of the working muscle by closing down the blood circulation to all other "collateral" or "unimportant" areas. By analogy one is entitled to suppose that the supply of protein to tissues is also governed by their demand; but if we accept this point of view, the question arises "who" in the body decides in times of shortage of external protein-supply which of the tissues are "important" and which tissues are "collateral" and can be sacrificed in reducing their bulk considerably.

There is another aspect which is just as intriguing, the problem of the so-called *protein sparers*. If enough carbohydrate is given to a fasting person, the daily N-excretion can be lowered to one-third of the "fasting" figure. If the caloric requirements are covered by carbohydrate and fat only, then the *minimum of nitrogenous excretion* occurs, but the subject under these conditions is *not* in nitrogenous equilibrium (negative N-balance) and soon reaches the stage of subclinical protein-calorie malnutrition. The *irreducible protein minimum* must be determined, with the nutritional cooperation of protein sparers, covering the caloric requirements of the body. This explains the low N-requirements found in the study of populations living to a large part on carbohydrates (e.g. Africa).

The action of protein sparers calls for a second question: "who" decides in the body that less protein should be used, if enough non-protein-calories from protein sparers are available? How is this homeostatic regulation performed?

The Vitamin and Enzyme Aspect

The discovery, isolation and synthesis of most vitamins has given rise to a "fashionable" line of research in nutrition, which obtained an important impulse when it became evident "that trace nutrients, such as vitamins, participate directly in enzyme function, either as precursors of co-enzymes or activators of enzyme activity. This hypothesis has been very successful in its application to the B-complex vitamins and trace minerals, but less successful in relationship to the fat-soluble vitamins and ascorbic acid" (Olson, 1967).

In malnutrition or hunger, two aspects with regard to enzyme function must be separated: (1) a diet can be deficient in certain vitamins, which are necessary for the body as co-enzymes; (2) a diet can be deficient in proteins, which are necessary for the building up of apo-enzymes in the body. Therefore, a loss of activity of an enzyme in malnutrition or hunger can be due to a deficit of co-enzymes or a protein deficit or both together. But it also happens that an enzyme activity first decreases and then increases. Such is the case for the disaccharidases (lactase, maltase, palatinase, sucrase) in the jejunal mucosa of rats, kept on a protein-free diet (Solimano, E. Ann Burgess and Levin, 1967). This behaviour shows that first there is a deviation from homeostasis due to lack of the protein necessary for building up the apo-enzyme, but then other homeostatic factors come into play, adapting the activity of these disaccharidases to the high carbohydrate diet, by diverting protein of other sources to cover the needs of this essential enzyme synthesis. Olson (1967) has shown that there is a third possibility: the fat-soluble vitamins A, D, K and E may have a control function in the protein synthesis for specific apo-enzymes. There is good evidence that vitamin K and E act on the genetic level of DNA-dependent RNA-synthesis by influencing transcriptional and translational processes. Nirenberg (1965) suggested that even "codeword" modifications could serve as factors with a regulatory or operator function in protein synthesis. Vitamin deficiency as such, protein deficiency as such, or both together are the cause of a serious disturbance to the enzyme-homeostasis in the body. The activity of certain enzymes is reduced, that of others is maintained, and in some the activity increases under the influence of this external stress. Rather selective changes occur of which we only begin to know something and which are a challenge to physiologists, biochemists and nutritionists.

The Protein Matrix, what is it?

We talk about the fluid matrix of the body and modern methods have made it possible to follow accurately any variations in the three compartments of this matrix (blood plasma, extracellular and intracellular space). The more we know about the proteins in our body, the more it becomes evident that the "protein matrix" must have a much more complex compartmental structure than the fluid matrix. Take for instance the plasma proteins of blood: albumin, even under conditions of protein deficiency in the food, is maintained at a fairly constant level, although other protein stores are already depleted. Why? Albumin is important for the onkotic pressure of the blood and thus for the equilibrium of water between blood and tissues. A fall of the albumin level below 3.5% leads to oedema —

proof of a serious failure of a homeostatic regulation. Gamma-globulin on the other hand is always higher in kwashiorkor than in healthy children, the alpha$_1$- and alpha$_2$-globulins and the beta-globulins are only faintly lowered (Sandstead et al., 1965). This example shows the selectivity of the body's response to an external protein deficiency.

Whipple (C. H. Whipple, 1948) divided the protein matrix into three compartments: (1) *fixed* cell protein, which is indispensable for cell life or activity; (2) *dispensable* reserve protein, which can be called upon for energy transformation or other purposes; (3) *labile* reserve protein, which can be readily turned out into the blood stream to maintain the concentration of plasma proteins, mainly albumin. This is a first approximation to the problem.

The Antagonism between Liver and Muscle

Rats can adapt to a reduced external protein supply by an alteration in the pattern of protein synthesis (Waterlow and Stephen, 1966). New protein is concentrated in the more vital tissues such as liver at the expense of the less important ones like muscle (Waterlow, 1959; Bendicenti, Mariani, Paolucci and Spadoni, 1959). This protein synthesis is due to an increase of the amino-acid activating enzymes in the liver, the synthetases (about twofold), whereas a steady state of the activity of these enzymes prevails in the heart or muscle (Mariani, Spadoni and Tomassi, 1963; Gaetani, Paolucci, Spadoni and Tomassi, 1964). If the diet contains protein sparers, the level of the urea cycle enzymes falls, leading to a decrease in urea excretion (Schimke, 1962). Protein synthetases and urea enzymes are two groups of enzymes which come into play as homeostatic factors acting in opposite direction, in order to adapt the body to a low level of protein intake. Amino acids reaching the liver are actively built up into protein by the amino-acid activating enzymes and protected from degradation into urea by the low level of the urea cycle enzymes (Waterlow, Alleyne, Chan, Garrow, Hay, James, Picou and Stephen, 1966). In these studies it became apparent that a very good "protein reference" can be obtained by measuring simultaneously the level of DNA-protein. In young rats the amino-acid activating enzymes and the total DNA content increase with the increase in weight of liver; the enzyme level per unit DNA remained, therefore, unaffected by growth. Under protein lack, on the other hand, the enzyme levels per unit DNA-protein were doubled in the liver, but only little change was found in muscle (Stephen, 1968). These findings were confirmed by biopsies on children under protein-calorie malnutrition and showed that in man the level of protein intake governs the activity of the liver enzymes regulating the internal protein supply (Stephen and Waterlow, 1968). It is obvious that with protein lack the protein syn-

thesis shifts from muscle to the liver. Returning to a normal diet, the
shift occurs in the opposite direction (Mendes and Waterlow, 1958).
Cortisone stimulates the incorporation of amino acids into liver (Munro,
1964) blood cortisol levels are therefore high in children with protein
malnutrition (Alleyne and Young, 1967).

These examples are not quoted in order to review all the recent work
on enzyme levels under the external stress of protein deficiency. They are
meant to serve as examples of how powerful homeostatic factors intervene
in order to protect the functional integrity of the protein matrix.

Attempt to classify the Enzyme Reactions in Protein Deficiency

As a first rough approximation one can distinguish three different
reactions: (1) maintenance of the normal level of activity of an enzyme,
opposed to the general depletion of protein reserves. This is a homeostatic
reaction favouring the "important" enzymes, such as for instance the respiratory enzymes; (2) decrease of the level of activity. Such a decrease can
have various causes: (a) shift of the site of protein synthesis to other more
important organs (e.g. from muscle to liver); (b) reduction of the protein
metabolism (e.g. influence of protein sparers); (c) shift in the channels of
the protein metabolism (such a shift is reflected in the change of the composition of urine with regard to products of protein metabolism);
(d) lack of apo-enzyme and other causes (e.g. vitamin deficiency); (3) increase of the activity of certain enzymes. This is a typical compensation
reaction, brought about by homeostatic factors, as a response to protein
deficiency. Enzymes of the carbohydrate- and fat-cycles fall into this
category, but also amino-acid activating enzymes.

Are we on the Right Way?

Walter Cannon considered homeostasis as the maintenance of bodily
equilibria at rest, under severe exercise and also in all cases of emergency,
such as deficient coagulation of the blood, low blood pressure, insufficient
water, excess water, insufficient or excess salt, low blood sugar, reduction
of blood and tissue proteins, low or excess concentration of calcium,
deficient oxygen, changes of pH, excess temperature, irritants, viruses
and bacteria, danger of death, etc. We are faced today with the acute
danger of famine for hundreds of millions of human beings, and of
protein-calorie malnutrition in infants, indeed an emergency which
should take the first priority in all our thinking and activity!

Which are the Merits of a Refined Study of Biochemical Data in the Battle against Protein-Calorie Malnutrition?

As an introduction to this chapter I would like to quote a few significant sentences from the "Report of the President's Science Advisory

Committee" on the "World Food Problem" (The White House 1967) with the understanding that the aspect with which we are concerned is only one little building-stone in a huge building presented as "World Food Problem".

"The developing countries should be encouraged and assisted in increasing and improving protein supplies in the following ways if the specific country has the natural resources and economic means." (B 2 p. 26) "Develop high-protein foods, especially for preschool children, from indigenous products." "Develop and utilise new genetic strains of plants that provide protein with a high nutritional value." "For each developing country or region, the extent and causes of protein-calorie malnutrition, the extent of deficiency of vitamins and minerals, the etiology of malnutrition, the relationships of nutrition to health and mental development, and the factors affecting acceptability of foods and food habits, will have to be determined."

These words may be considered as the basis of the program of *Nestlé Foundation*'s work in the Ivory Coast. And now the question arises, which methods are available to determine the extent and causes of protein-calorie malnutrition, the extent of deficiency of vitamins and minerals? How can one guide the agricultural experts to develop and utilize new genetic strains of plants that provide protein with a higher nutritional value?

Up to now, as far as I know, a line was drawn between "scientific field work" and research in the hospitals of developing countries. In other words, out in the field only very primitive and rather old-fashioned methods could be used. In the hospitals, on the other hand, modern biochemical methods are available but they were mainly applied to study severe clinical cases of protein-calorie malnutrition, or in other words: patients were examined who had already lost their power of homeostatic regulation and usually had other infectious complications. Physiological protein-calorie malnutrition is found in communities out in the field, where the subjects are still "healthy", due to a homeostatic regulation maintaining desperately the function of the protein matrix against the constant stress of an insufficient N-supply. These subjects will respond favourably to an increase of proteins in their nutrition.

Test proteins of equal N-content can be studied with regard to their "curative effect" on the biochemical pattern of enzymes and nutrients in blood, liver and muscular tissue, e.g. modern biochemical methods are so accurate and sensitive that very extensive studies can now be made on small samples of blood and tissues obtained by biopsy. The power of a protein in restoring the homeostatic equilibrium should be considered as its nutritional value per gram of N. Guided by such modern biochemical work, plants which provide proteins with a higher nutritional value can be classified and selected with regard to their nutritional value. The next step is a study of their acceptability by the population with regard to agricultural and food habits. Based on such a detailed biochemical study, a pioneer experiment with a larger community can be started and this may have a decisive influence on the nutritional policy of a whole country.

A refined biochemical study of a rural population has the following advantages: (1) in such a population relative uniform food habits prevail; (2) a much closer contact with the population can be established than in cities or large agglomerations; (3) protein-calorie malnutrition is "endemic" in rural populations; it is here where "sanitation" is needed! There are of course also disadvantages: (1) the gastrointestinal infection rate with parasites is high; (2) food habits and taboos are opposed to new ways of nutrition. We hope that our plan to start such a study in the Ivory Coast will be successful.

Homeostasis and Nutritional Requirement

All cells of our body require a more or less wide range of supply of materials for their growth and survival. Amino acids, mineral elements and vitamins, the "nutrients" are essential for growth and maintenance of the functions of the cell. Proteins, carbohydrates and fats are sources of energy transformation. Limiting our attention to the maintenance of the protein matrix of the body we can distinguish: (1) optimal protein intake (optimal nutrition); (2) sub-optimal protein intake, balanced by homeostatic reactions of the body (sub-optimal nutrition); (3) sub-minimal protein intake, leading to a deficiency disease.

From the homeostatic point of view, a "time reserve" has to be considered in conditions of sub-minimal protein intake. A negative N-balance can be regulated by homeostatic factors as long as internal protein reserves are available, in other words, the duration of homeostasis under the external stress of a sub-minimal intake of protein is dependent on the availability of internal reserves and the efficiency of the homeostatic reactions. If the reserves are depleted, homeostasis cannot be maintained, the physiological range of adaptation comes to an end and this is the moment when the *deficiency disease* begins (kwashiorkor, marasmus, pellagra, etc.). In starving experiments of healthy persons the "time reserve" can last up to 4 weeks (Benedict, 1915); if a sub-minimal protein intake is combined with an optimal intake of protein sparers, the "time reserve" may extend over very long periods. But the homeostatic reactions are only able to maintain the equilibrium by reducing not only the metabolic rate of the body, but also the mental and physical activity. In animal experiments it became evident that during the growth period a sub-optimal protein intake produces damage which cannot be repaired by a later return to an optimal nutrition. This is a very serious point, showing that homeostatic emergency reactions may lead to permanent damage, if they are pushed to the limits of physiological regulation.

How is the Homeostatic Regulation of the Protein Matrix brought about?

We know very little about the regulating centers and even less about the signals which inform them of the peripheral requirements and which carry the orders into the tissues. In the hypothalamus a "feeding center" and a "satiety center" have been localised (see Anand, 1961). The cerebral structures of the frontal and temporal lobes included in the limbic system direct what might be called "discriminative appetite" and the highest level of the brain, the neocortex, governs feeding habits and conditioned reflexes. With regard to the afferent and efferent signals, various hypotheses have been discussed: (1) thermostatic regulation; (2) glucostatic regulation; (3) lipostatic regulation; (4) regulation by plasma amino acids; (5) regulation by afferent impulses from the presence of food; (6) regulation by correlation of water and food intake. No single hypothesis can be considered as satisfactory and, if we know more about this important homeostatic regulation, a multiple factor mechanism will be revealed.

In looking back at Cannon's 6 postulates it seems to me that they still provide a considerable stimulus for the awareness of our ignorance about the regulation of the functional integrity of the protein matrix of our body.

From the Nutrition Unit, World Health Organisation, Geneva (Switzerland)

Outline of WHO and PAG Participation in Research on Protein-Calorie Malnutrition

J. M. Bengoa

With 2 Figures

Contents

WHO, in selecting the fields in research as well as in applied nutrition, has taken into account some criteria for establishing priorities. The criteria have been based on the following factors:

   1. Extent and magnitude of the problem, judged by its:

      a) mortality,

      b) morbidity, and

      c) irreversible damage.

2. Effect on productivity.

3. Feasibility of prevention.

4. Interest of authorities.

An analysis of these factors was made in a paper presented to the Seventh International Congress of Nutrition in 1966, in Hamburg (Bengoa, 1966).

Based on the above criteria the following four conditions were considered to have a high priority, in the present-day context:

a) *Protein-calorie malnutrition*, or protein-calorie deficiency diseases in young children.

b) *Xerophthalmia*.

c) *Nutritional anemias*.

d) *Endemic goitre*.

Other nutritional conditions may have a higher priority in certain limited areas but, considering the problem on a world-wide basis, the above-mentioned stand out as the most important.

The resources available from the WHO regular program in support of nutrition research amount to approximately $150,000 a year, of which two thirds are allocated to research on protein-calorie malnutrition, including human testing of the new protein foods. Besides this, WHO provides a variable amount of further assistance to individual workers through grants from its general research co-ordination division. Mention should also be made of research activities undertaken by PAHO and its institutes with its own funds and funds provided by foundations such as NIH, Williams-Waterman Foundation and others.

A good part of this research is being carried out by well-known research workers in different countries, chosen by the merits of the proposal and its compliance with WHO priorities. Special mention should be made of the role played by INCAP in nutrition research, and the results of some of its studies on protein-calorie malnutrition and protein-rich mixtures are well known to most of you. The PAG, however, does not carry out research by itself, but its indications, guidance and recommendations are of great value to the three sponsoring organisations, namely, FAO, WHO and UNICEF.

Following the epidemiological approach, the main research activities of WHO, HQ on protein-calorie malnutrition, are problems related to: (1) the host; (2) the agent; (3) the environment.

1. *Research on the host:* One of the important aspects to elucidate in this area is the question of precise protein needs of infants and young children as well as pregnant and lactating women. The results of studies carried out and their implications are given below briefly.

a) Protein requirements of infants in Jamaica, by Chan and Waterlow (1965, 1966): The requirements were measured in terms of milk protein.

The average requirement for maintenance was found to be 100/120 mg nitrogen per kg per day. Excellent growth was obtained on intakes of 200 mg. These results, apart from providing basic information on the infant's needs, served to determine the needs in terms of milk substitutes in the weaning period.

b) Protein requirements of pre-school children: This investigation, which started in 1967, is being carried out in Guatemala (INCAP), by Dr. Arroyave. The project proposes to study the biological behaviour of children of pre-school age in relation to the protein intake, fed at various levels. Measurements of hydroxyproline, as collected in urine, was selected as the index of protein nutrition status. These studies will perhaps be presented by the author in one of the sessions.

c) Protein requirements of pregnant women: This investigation has been carried out in Ibadan, Nigeria, by Professor Edozien. Two methods of approach were used. The first one was to determine the minimum level of dietary protein required by lactating women to maintain a satisfactory state of protein nutrition, as measured by the blood hemoglobin level and the concentration of proteins in the plasma; maintain nitrogen equilibrium; and produce enough breast-milk to ensure a satisfactory rate of weight increment of the infant.

The second procedure was to determine the protein intake of the breast-fed infant and to take this as an estimate of the additional protein which must be made available to the mother to compensate for the protein secreted in the breast milk. Although valuable results were obtained, it is unfortunate that the study could not be completed owing to local circumstances.

Needless to say, the above studies contributed a good deal to our understanding of the needs of an important nutrient in the vulnerable phases of life. Based on the results of these and similar studies, FAO and WHO/FAO have brought out reports on protein requirements (FAO, 1957; WHO, 1965).

2. *Research on the agent:* that is to say, on the possibility of increasing the availability of proteins, particularly those products directed to the weaning period.

Our main responsibility with reference to products developed under the auspices of international organisations is obviously to carry out tests in human beings on the utility, acceptability and tolerance before a given product is authorised for human consumption.

A guide prepared by WHO and endorsed later by PAG provides the details on the conditions in which the human testing has to be developed. DeMaeyer (to be published) has recently summarised the guide and the results of human testing.

Four types of tests recommended by PAG have been utilised in some of the studies so far carried out. They are:

a) Acceptability and tolerance, particularly in young children.

b) Growth test.

c) Nitrogen balance.

d) Biochemical tests.

Institutes of five countries in the world [Ethiopia, India, Chile, Guatemala and China (Taiwan)] participated in these studies. The table (on p. 14) gives a summary of the recent products which WHO and PAG have under consideration.

While some initiative and forevision were necessary in evolving these products suitable to the local conditions, the acceptability and tolerance tests had shown that all of them can be used satisfactorily to prevent or ameliorate the problem of protein-calorie malnutrition. However, the difficulties that are still to be faced are in the fields of commercial production and particularly the marketing of the products tested. It is here that more research is needed with the creative imagination of sociologists, economists, marketing experts and food technologists.

Our main task in WHO is to motivate the medical and para-medical workers on the value of the new formulas since, without their support, it will be difficult for the community to accept a new product. An important approach in this context lies in the possibility of increasing the protein value of commercial products already in existence and well recognised in the market, rather than giving new names to the new products.

From my experience as a rural doctor 30 years ago, in Venezuela, I always remember a product called Nenerina, widely used by mothers even in the remote areas. The protein content was very low and it was based on cereals. I do not consider it to be extremely difficult to add something else to this product without changing its colour and flavour but increasing its protein value to 15 or 20%. The new product, however should be sold *with the previous name*, as Nenerina. In areas where there are already weaning products based on cereals, it is perhaps better to take advantage of their acceptance in the market and improve the quality instead of injecting into the population new names and new products. In some countries an attempt has been made to legislate the composition of weaning foods, fixing a minimum of protein. Some subsidies from the government may be necessary in order to avoid any increase in the price of the product, until such time as the product gains a wider demand.

3. *Research on environmental factors:* Among the various factors, the infectious diseases, which are well known to be the conditioning or precipitating factors of malnutrition, have been the subject of WHO research in recent years. A study undertaken in South Africa in 1964 demonstrated an inseparable relationship between child mortality rate

Table 1. *Recent formulas developed for possible use in international programs*

Wheat-based protein-rich mixtures: Composition in percentage

	Tunisia	Algeria	UAR	Turkey	66/10/202	66/10/201
Wheat, %	31	28	25	40	32 (whole)	52 (whole)
Chickpeas	34	38	25	20	30	—
Lentils	26	18.5	25	—	—	—
Soya flour	—	—	—	20	30 (full fat)	40 (full fat)
Skim milk	—	10	5	10	—	—
Sugar	8	5	18	8	7	7
Vitamins, minerals, flavour	1	0.5	2	2	1	1

Corn-based protein-rich mixtures: Composition in percentage

	CSM	66/10/203	Mx-49	Mx-50	Mx-51	Mx-52
Corn, %	68	43	40	58.5	41	48
Soya flour	25 (defatted)	40 (full fat)	38 (full fat)	35 (full fat)	42 (full fat)	30 (full fat)
Skim milk	5	—	5	5	—	—
Fish protein concentrate	—	—	—	—	—	5
Sugar	—	15.5 (white 10, raw 5.5)	15.5	—	15.5	15.5
Vitamins, minerals, flavour	2	1.5 (iodised salt 0.5)	1.5	1.5	1.5	1.5

Miscellaneous:

	Mx-37	Mx-39
	$\frac{1}{3}$ soya (full fat)	rolled oats 85%
	$\frac{2}{3}$ whey	soya flakes 15%

especially from diarrhoea, and malnutrition (Wittmann, Moodie, Felling-ham and Hansen, 1967); another study supported by WHO in India proposes to investigate the effects on the growth and development of pre-

Table 2. *Mortality from gastro-enteritis, measles and whooping-cough, for children under 5 years*[a]

	Gastro-enteritis, 1964[b]		Measles, average 1963-64		Whooping-cough, 1964	
	under 1 year rate per 100000 live-born	1—4 years rate per 100000 population	under 1 year rate per 100000 live-born	1—4 years rate per 100000 population	under 1 year rate per 100000 live-born	1—4 years rate per 100000 population
Austria	139.0	6.8	4.5	2.2	2.2	1.0
Denmark	40.8	1.7	—	1.0	2.4	0.3
Sweden	13.0	1.4	—	—	—	—
United Kingdom: England and Wales	42.0	3.5	2.1	1.5	3.7	0.3
Chile	1588.0	69.3	413.2	137.6	68.3	8.1
Colombia	1480.4	278.6	61.9	48.8	191.2	52.8
Guatemala	1275.9[c]	704.2[c]	248.0[d]	310.0[d]	589.4[c]	309.2[c]
Mauritius	1380.3	203.2	1.8	—	3.6	—
Mexico	1224.0	267.4	71.0	103.7	112.7	75.8
Philippines	521.0	149.6	58.4	25.6	12.3	2.6
Venezuela	714.7	89.8	24.2	21.9	53.6	15.2

[a] From National and WHO International Reports.
[b] Gastritis, duodenitis, enteritis, and colitis, except diarrhoea of the newborn (B36).
[c] 1963.
[d] Average 1962—1963.

school children as a result of control of infections; yet another study in Ghana pays special attention to infectious diseases that precede the onset of malnutrition.

Since the importance of the interrelationship between nutrition and infection is of great concern to WHO, a comprehensive monograph on this subject has been brought out last year (Scrimshaw, Taylor and Gordon, 1968).

The seriousness of this problem can be best illustrated by the history of a Latin American child, taken from INCAP, which is by no means exceptional (Fig. 1). During the first 2 years of his life, this child passed through 6 episodes of conjunctivitis, 5 of diarrhoea, 10 of upper respiratory disease, 4 of bronchitis, 1 of measles followed by broncho-pneumonia and 1 of stomatitis. On the whole, 28 attacks of infectious diseases were encountered in the child in 24 months. He was ill for 29% of his life, (Mata, Urrutia and Gordon, 1967) and probably convalescing for another 30% of the time.

WHO has very limited resources available to cover many other aspects related to protein-calorie malnutrition. Much more has to be done in the three directions mentioned above, namely; on the host, particularly on children, without neglecting the problem of the adults; on the agent, decelerating the trend of early weaning, developing new sources of proteins, but also increasing the calorie supply, and the improvement of the whole diet; and on the environment, particularly the question of infectious diseases, socio-economic conditions, cultural and educational factors and so on, with a view to improving the currently unsatisfactory trends and practices in child feeding.

The results obtained by health activities all over the world in reducing the mortality rates during the last 20 or 30 years have been spectacular and are widely recognised (Fig. 2). An obvious but not thoroughly studied result of this success is that the number of survivors has considerably increased. Therefore, besides the continuing efforts to decrease the mortality rates still further — for there is still a difference as compared with the more advanced countries — attention needs to be paid to the present and future situation of these survivors. Survivors who would probably have died 30 years ago are saved today, thanks to public health action and also to the concomitant socio-economic progress. They escape death, but how many can be regarded as having been completely restored, both biologically and socially, and how many remain with lasting damage and are mere survivors ?

As I do not wish to encroach upon your valuable time, I will not elaborate on the WHO research efforts in the other fields of xerophthalmia, anemias and goitre, referred to earlier. It will be sufficient to say that similar efforts, with the main objective of finding feasible solutions to the pressing problems, are also in progress with reference to the other disease conditions.

I wish, however, to take this opportunity to quote a statement made in the Seventh Report of the Joint FAO/WHO Expert Committee on

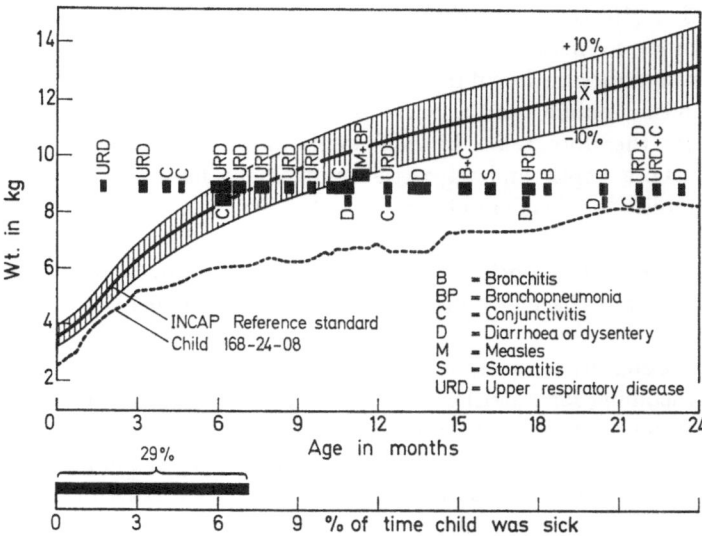

Fig. 1. Weight and infectious diseases in a Guatemalan village child

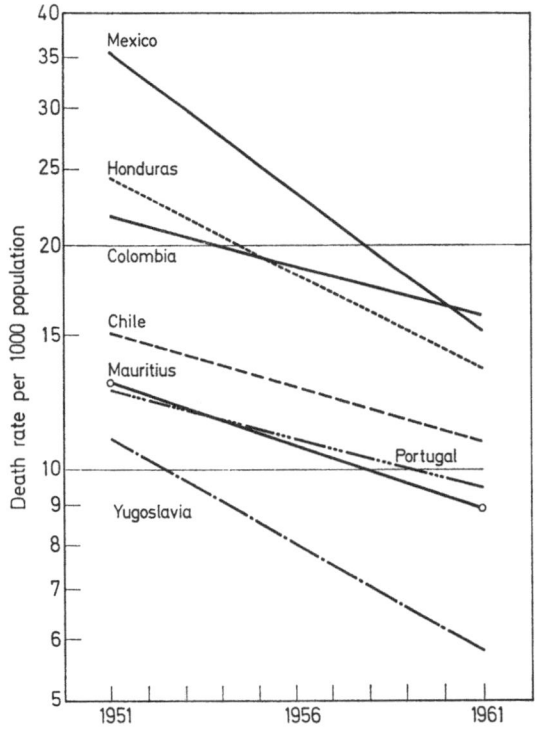

Fig. 2. Trends in the mortality rate of children 1 to 4 years of age (1951 to 1961)

Nutrition (1967) in the section related to research in nutrition: "The nutritional scientist is not expert in all fields related to production, processing, distribution of foods, development of attitudes towards foods, metabolism, public health, behavioral sciences and population control. He must, therefore, relate himself closely in research to scholars in other fields whose depth and expertise complement his own. He must not oversimplify his problem of experimental design in an effort to identify a single cause or risk factor in a given situation. This consideration becomes especially important as nutrition research moves more and more into studies of the many diseases which are but tangentially related to diet.

The Committee wishes to express its concern that adequate support be forthcoming for nutrition research and urges that foundations, other research funding agencies, and national, bilateral and international agencies include realistic support for research activities in the broad field of food and nutritional sciences."

From the Department of Biochemistry of the University of Berne (Switzerland)

Enzymes and Nutrition

H. Aebi

With 12 Figures

Contents

Introduction

This contribution reports on the progress made in the borderland between enzymology and nutrition; in particular a few problems relevant for the study of malnutrition will be discussed.

The level of a particular enzyme in a certain type of tissue is governed by many factors, such as inheritance, age, sex, environment and diet. Therefore, the enzymatic activity in any sample of tissue or body fluid represents a complex resultant of a multitude of synergistic and antagonistic effects. A change in diet may affect the activity of an enzyme in different ways. There are at least three possibilities by which such an alteration can be accomplished (Hess and Brand, 1965).

Mechanisms of Enzyme Regulation

1. Allosteric Inhibition or Activation

Slight changes in the conformation of an enzyme molecule may lead to a drastic increase or decrease in specific activity and affinity for a particular substrate. This is mainly observed in enzymes composed of two or more sub-units. Either a precursor substance or a metabolite may indirectly affect the properties of the substrate binding site by acting on the group which specifically binds the regulator substance. Allosteric inhibition represents a mechanism by which the overall rate of an entire reaction sequence can be controlled (Gerhard and Pardee, 1962). This is an immediate response to a change in the metabolite pattern operating in many regulatory feed-back systems (e.g. regulation of aspartate-transcarbamylase by cytidine triphosphate).

2. Alteration of the Enzyme Concentration

The adjustment of the level of enzyme concentration is a slower but very efficient mechanism due to a change in the rate of synthesis or degradation. Since the synthesis of all proteins is under genetic control, it is evident that genetic regulatory mechanisms, such as induction or repression, play an essential role in continuously adapting the metabolising capacity of the organism to the actual environmental conditions. Not only in micro-organisms, but also in mammals, many enzymes are regulated efficiently by inducers (e.g. tyrosine-inducing tyrosine-transaminase or tryptophan-inducing tryptophan pyrrolase).

3. Limitation of Enzyme Formation by the Availability of its Constituents

If an adequate supply of amino acids and (or) coenzymes cannot be maintained at the site of enzyme synthesis, it is evident that their restricted availability or a selection among various messages obtained may become a controlling factor in the synthesis of an enzyme. Recent studies by Munro (1964, 1968) performed on rat liver polysomes in vitro have shown that the functional state of the protein synthetising mechanism is dictated by the supply of amino acids as to quantity and composition. The incorporation rate of labelled amino acids (e.g. ^{14}C-leucine) is adapted within 1 min. This change can be visualised by an alteration in the polysome profile obtained by centrifugation of the microsomal fraction in a sucrose gradient. If a complete set of amino acids is added to the system, polysomes are almost exclusively found, indicating that protein synthesis proceeds at full speed. However, if no amino acids are present or if an incomplete set of amino acids is added (e.g. lacking tryptophan), there is no synthesis or it is stopped at an early stage. Accordingly ribosomes are mainly present as monomer or oligomer. Therefore, what has been known at the individual level for a long time in regard to amino-acid requirements for promoting growth can be studied today at the molecular level.

As shown in the lecture by A. von Muralt, homeostatic mechanisms operate at every level in order to maintain optimal intracellular working conditions. To reach this aim, all mechanisms mentioned above act or interact simultaneously. Considering enzymes to be something like molecular machines, the body can accomplish adaptations by producing more machines or by forcing them to turn at a higher or lower speed or eventually by switching them off. It is evident, however, that these regulatory mechanisms can operate under favorable conditions only as long as a critical limit to the availability of amino acids has not been exceeded.

Specificity of Dietary Effects on Enzyme Levels

The study of dietary factors affecting enzyme activity has revealed the following features.

1. Organ Specificity

There is a characteristic organ specificity. In most instances liver tissue exerts more pronounced alterations in quantity as well as quality than is the case in muscle or other organs. Since the response of liver to a change in the diet is usually quite significant and well reproducible, its composition and the concentration of enzymes in liver have

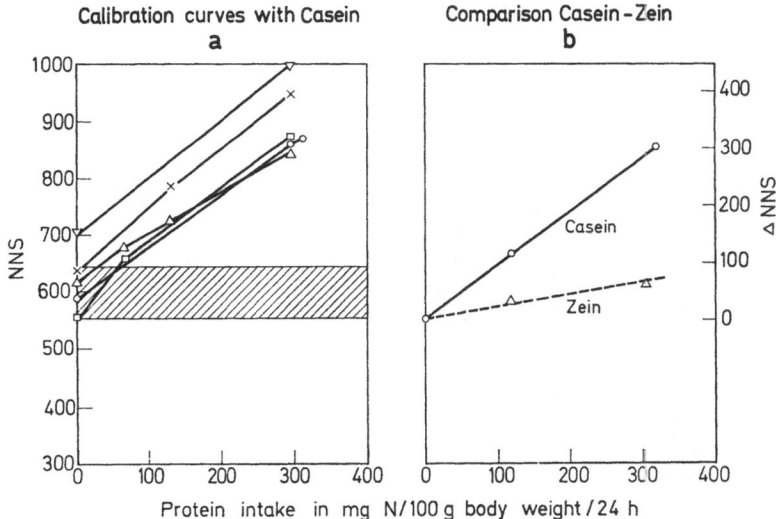

Fig. 1. Total protein content of rat liver as a function of protein intake. NNS = non-lipid, non-glycogen solids. Left side: Calibration curves obtained by variation of the casein content of the diet. Right side: Comparison between casein and zein. (Aebi, 1964)

often been used as an indicator for the evaluation of the quality of a diet. Kosterlitz (1944) has determined the nutritive value of a protein by feeding it to rats for a period of 4 to 6 days and subsequently analysing the liver for its total protein content or by simply measuring the NNS (= non-lipid, non-glycogen-solids) (Fig. 1). Miller (1948) was one of the first authors who studied the effect of dietary factors on liver enzymes. Based on his observations, Williams and Elvehjem (1949) introduced a new technique for the estimation of the biological value of a protein by using xanthine oxidase activity in liver as an indicator.

2. Specificity of Response

Specificity of response of different enzymes: From the systematic study made by Wainio et al. (1959), it can be seen that different categories of liver enzymes seem to exist: those which undergo only a slight reduction if the rat is fed on a protein-free diet (e.g. cytochrome oxidase) and those which are drastically reduced in activity (e.g. uricase and xanthine oxidase). Furthermore, there is a difference in response if the quantity of food is reduced or if protein alone is withdrawn. This is most pronounced in D-amino acid oxidase. Although there is no proof, these differences have to be considered as an expression of non-proportional utilisation of the amino acids available. Therefore, it may be assumed that in any manifested state of protein deficiency certain priorities are respected (Fig. 2).

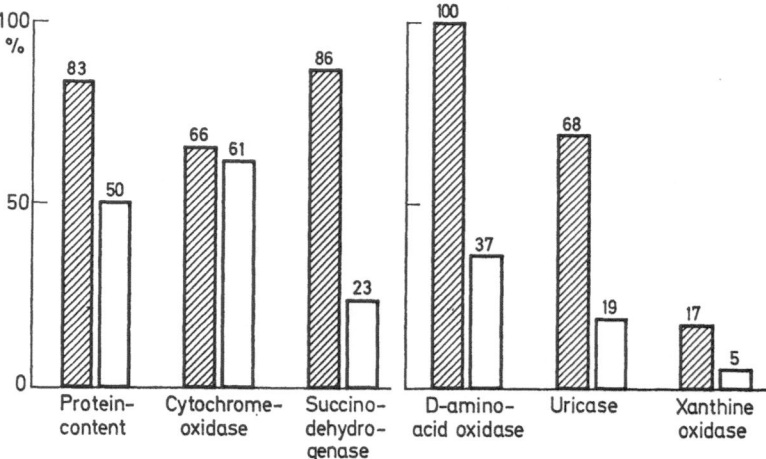

Fig. 2. Dietary effects on protein content and enzyme levels in rat liver. ▨ = 25% restriction of total food intake (standard diet), □ = Protein-free diet. All data are related to those of control animals fed ad libitum on a standard diet (18% casein). (Wainio et al., 1959)

3. Coordination

Coordination in the regulation of enzymes forming a team: Whenever the metabolic capacity of a cycle or a sequence of reactions is altered, adaptation is accomplished by a proportional increase (or decrease) in activity level of all enzymes participating in the reaction. There are two classical examples which impress by the extent and the specificity of the effects observed:

a) Schimke (1962) has shown that the total liver content of all five urea cycle enzymes, i.e. carbamyl-phosphate synthetase, ornithine

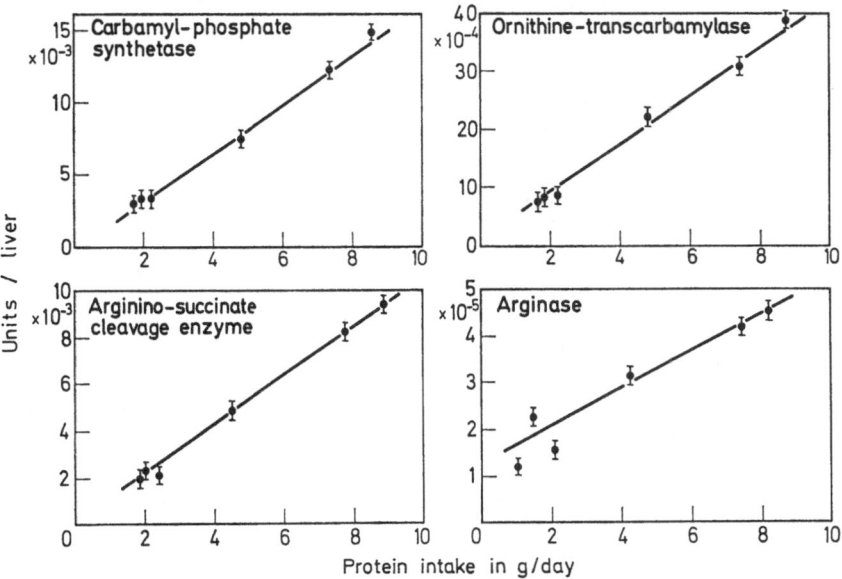

Fig. 3. Relationships of total liver content of urea cycle enzymes to daily protein consumption. Protein intake in grams per day is plotted against enzyme content in units per liver. Brackets indicate ± 4 standard errors. (Schimke, 1962)

transcarbamylase, argininosuccinate cleavage enzyme, argininosuccinate synthetase and arginase are directly proportional to the daily consumption of protein (Fig. 3). However, lactate dehydrogenase and glucose-6-phosphate dehydrogenase exerted a different pattern of change, whereas malic and glutamic dehydrogenases were not affected at all by a change in the protein intake.

b) Fitch and Chaikoff (1960) as well as Weber et al. (1961) have reported that in rats fed on a diet high in fructose (60%) for 7 days there is a considerable increase in activity of many enzymes involved in sugar metabolism, compared with rats fed on a standard diet containing starch as the main carbohydrate. This effect is particularly striking in enzymes

producing NADPH (glucose-6-phosphate dehydrogenase, 6-phospho-
gluconate dehydrogenase) or consuming NADPH (malic enzyme),
where the increase is almost tenfold (Aebi and Richterich, 1963; Fitch
and Chaikoff, 1960) (Fig. 4).

Glycolysis, however, is the best example to demonstrate this principle
of coordinated activity regulation in a team of enzymes (Garfinkel and
Hess, 1964). As shown by Bücher (1962) and his collaborators (Pette and

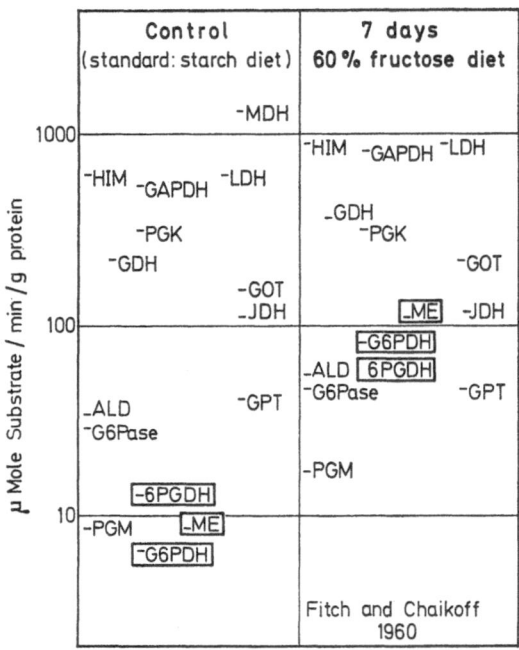

Fig. 4. Effect of a diet high in fructose (60%) on the enzyme pattern of rat liver. A
tenfold increase is observed in a) glucose-6-phosphate-dehydrogenase (G-6-PDH),
b) 6-phosphogluconate-dehydrogenase (6-PGDH) and c) malic enzyme (ME).
(Fitch and Chaikoff, 1960)

Bücher, 1963), this is particularly true for the branching-free part of the
glycolytic chain, i.e. for a team of five enzymes, which — due to lack of
alternatives — always cooperate in a sequence. They are: triose-phosphate
isomerase, glyceraldehyde-phosphate dehydrogenase, glycerate-3-phos-
phate kinase, glycerate-phosphate mutase and enolase. Whatever type
of comparison is made (e.g. between organs, species or dietary effects),
there is always a synchronous change in the activity level so that all proportions among the five enzymes remain constant (Fig. 5). The existence
of such groups of enzymes exerting constant proportions may be taken

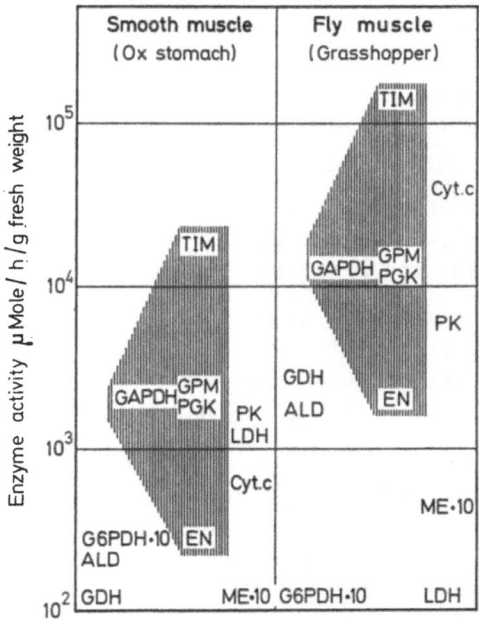

Fig. 5. Enzyme pattern in muscle tissue of different origin. The 5 enzymes of the branching-free part of the glycolytic chain are located in the triangle-shaped figures. (Bücher, 1962)

as evidence that also in the mammalian organism the level of enzyme activity and its adaptive alterations are subject to a central control.

Dynamic Aspects of Enzyme Regulation

The interplay of both enzyme formation and degradation in the control of enzyme levels has been studied in considerable detail by Schimke (1962, 1966). He has shown that in mammalian tissue, particularly in liver, there is an extensive continual synthesis and breakdown of total protein as well as of specific enzymes. Consequently an observed enzyme concentration represents the resultant of both, the rate of synthesis and the rate of degradation. Furthermore he has shown that either process can be altered independently by changes in the physiological and nutritional state by the administration of hormones or by applying inducers. The determination of the turnover of an enzyme permits the calculation of its half-life time. There is a marked heterogeneity of turnover rates among enzymes present in rat liver. The half-lives vary within a range of two decades. The half-life time of the following enzymes has been determined: ALA-synthetase[1] t/2 $\sim$ 1 h, tryptophan pyrrolase 4 h, catalase 24 h and arginase 4 to 5 days. The mitochondrial stroma proteins have an even higher half-life of about 7 days (Schimke, 1968). Therefore, the half-life time is specific for each enzyme and may vary within a wide range — even among species localised in the same compartment or organelle.

The relationship between the actual level of an enzyme (P) and its rates of synthesis (K_S) and of degradation (K_D = first order rate constant) can be expressed by the following formula:

$$\frac{dP}{dt} = K_S - K_D \cdot P$$

Thus in the steady state the equation becomes:

$$P = K_S/K_D$$

This highly simplified model offers an explanation, why — as a rule — enzymes present in low concentrations or exerting a high turnover are particularly sensitive to adaptive alterations of their level. Rapidly regenerating enzymes such as tryptophan pyrrolase or tyrosine transaminase respond faster and more strongly to cortisone administration than is the case for arginase.

The technique used for the analysis of enzyme turnover is based on the radioactive labelling of the enzyme under investigation. This is done by feeding a diet containing ^{14}C-lysine or ^{14}C-arginine until the level of specific activity in regard to ^{14}C incorporated into the liver proteins has reached a steady state (Fig. 6). This is the case in the rat after a feeding period of 3 to 4 weeks. Then the labelled amino acid is replaced in the diet by the non-labelled compound. Total enzyme activity and the ^{14}C activity of the isolated enzyme are then followed by analysing liver samples at appropriate intervals. By using this sophisticated procedure it has been shown that, in the case of tryptophan pyrrolase, glucocorticoid administration raises the enzyme level as a result of an increased rate of enzyme synthesis, whereas the substrate-induced accumulation of this enzyme has to be attributed to a complete stop of enzyme degradation, in presence of continued enzyme synthesis. From Fig. 7 it can be seen that total ^{14}C-activity in tryptophan pyrrolase is maintained at a constant level in presence of the inducer. However, there is a rapid fall in the control animal. Accordingly the level of the enzyme is increasing rapidly in the former and remaining constant in the latter.

Recent experiments performed by Schimke (1968) and his group have revealed that there are analogous changes in the synthesis-degradation interplay whenever the composition of the diet is altered. Considerable

---

[1] Amino levulinic acid-synthetase.

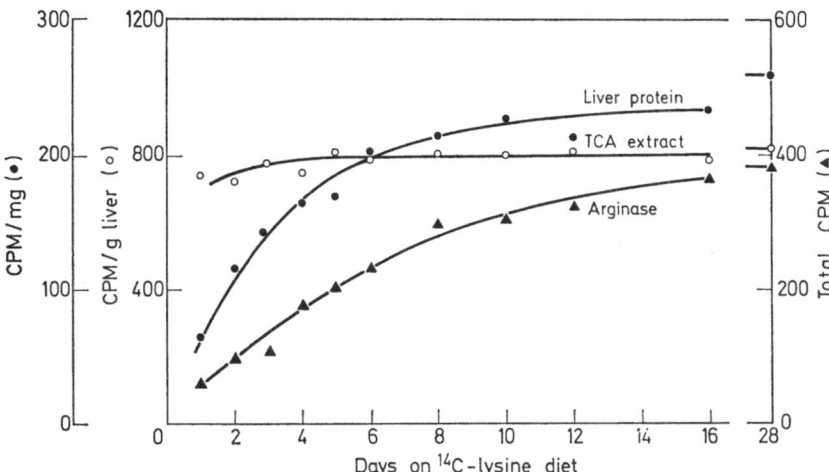

Fig. 6. Incorporation of continuously administered ¹⁴C-L-lysine into total protein, arginase and trichloroacetic acid (TCA) soluble extracts of rat liver. Osborne-Mendel rats maintained for 7 days on a diet consisting of 25% complete amino-acid mixture were then placed on a similar diet containing ¹⁴C-L-lysine. At intervals one rat was killed. Radioactivity of the TCA-soluble fraction is expressed as counts per min. per extract from 1 g of liver, wet weight (O). Counts in total liver protein: counts per min. per mg of protein (●). Counts in the arginase: total counts precipitated (▲). (Schimke, 1966)

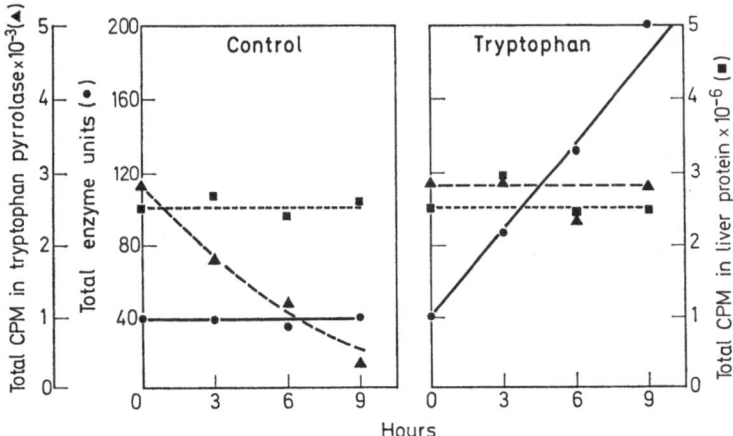

Fig. 7. Effect of L-tryptophan administration on the loss of tryptophan pyrrolase prelabelled with ¹⁴C-L-leucine. Rats were given single injections of 20 μC of ¹⁴C-L-leucine. Sixty min. later, two animals were killed. The remainder were given 10 ml of 0.85% NaCl or 10 ml of 0.85% NaCl containing 150 mg of L-tryptophan. These injections were repeated in the remaining animals at 4 and 8 h. The symbols used refer to enzyme activity (—●—); total radioactivity in protein precipitated by the tryptophan pyrrolase antiserum (- - ▲ - -); radioactivity in total cellular protein (- - ■ - -). (Schimke, 1966)

alterations in either rate do occur, e.g. after the level of casein has been lowered suddenly from 70% to 8%. Here the new steady state, with regard to arginase activity, is reached by both a decrease in the rate of synthesis and a reversible increase in the rate of degradation. However, if feeding is stopped entirely, there is an immediate and complete cessation of enzyme degradation. Here, for obvious reasons, well known to all who remember the economic situation in "austerity", the organism tends to save the enzyme. These observations are another good example of the effectiveness and the usefulness of homeostatic mechanisms. It is improbable that similar experiments can be accomplished in man. Nevertheless, the results of these investigations must be kept in mind whenever changes in enzyme levels have to be evaluated.

Some Factors modifying the Dietary Influence on Enzyme Levels

The complexity of a response to changes in environmental conditions makes it often very difficult to ascribe a certain effect to one single factor exclusively. In particular, there are close relations between nutritional and hormonal influences. The same is true for the interactions existing between oscillations due to intermittent food intake and circadian rhythms. Consequently such effects are always composed of two or more components. There is either an enhanced response because of the summation of single effects or they may compensate each other; thus in the latter instance an effect may be masked.

1. Hormonal Effects

Among the hormonal effects, above all the corticoids have to be considered. In many respects a high protein intake and doses of cortisol act in a synergistic manner. As shown by Freedland *et al.* (1968), this is true for the level of glutamate-pyruvate transaminase, glutamate-oxaloacetate transaminase and glucose-6-phosphatase activity in rat liver. Therefore, it is permitted to state that stress or application of cortisol doses imitate the effects seen when a protein-rich diet is fed and vice versa. This explains why the effect on the level of these enzymes, seen after a test-dose of cortisol, largely depends on the protein intake. In rats fed on a protein-free diet for 5 days, glutamate-pyruvate transaminase activity rises about tenfold. The effect — when the protein content is 25% — corresponds to a threefold increase. However, when a diet containing 90% protein is fed, only a slight increase (1.2 fold) can be observed after cortisol administration. In view of these data it is not unreasonable to assume that all factors exerting stress reactions may affect enzyme levels or other clinical-chemical data, especially when the individual is in a state of protein malnutrition. Other types of interdependence of hormonal effects and dietary factors have been observed between thyroid hormones and the level of protein and fructose intake. They all show that the type of response to hormonal effects is largely influenced by the nutritional status of the individual.

2. Feeding Schedule

Oscillation of enzyme levels: The role of the feeding schedule on the level of various enzymes in rat liver has recently been studied in detail by Potter (1968). By adapting rats to a controlled feeding schedule (i.e. 16 or 40 h fasting period and 8 h feeding period) and comparing them with animals fed ad libitum, he was able to demonstrate that, in a population of animals, the degree of variation can be reduced considerably. Owing to a high degree of homogeneity in his data, tidal changes in enzyme levels could be observed. This is particularly the case in tyrosine transaminase of rat liver. The rhythm to be observed shows two maxima situated at different times of the feeding cycle. The first maximum occurs at night and is considered as a "feeding peak" since it coincides with the feeding period. The second peak occurs late on the following day and is due to a circadian rhythm ("fasting peak"). The resulting bimodal curve thus reflects the controlled feeding schedule on which a circadian rhythm is superposed. So far, 12 enzymes have been shown to oscillate in a similar manner as described here for tyrosine transaminase. From this study it may be concluded that any intermittent feeding schedule may lead to tidal changes in the metabolite as well as the enzyme pattern. Therefore, it is not unreasonable to believe that, in a series of data obtained at various points of the feeding cycle, oscillation may significantly contribute to their heterogeneity.

Biochemical Individuality

The recognition that certain enzymes may exist in multiple molecular forms in the same individual is one of the most interesting developments in enzymology in recent years. Enzyme heterogeneity has first been observed in lactate dehydrogenase. The existence of various isoenzymes occurring in different tissues of body fluids in a certain proportion has opened new perspectives for the study of genetic control of enzyme formation and tissue differentiation. Since the isoenzyme pattern is different in most organs, the heterogeneity of enzymes also offers additional diagnostic possibilities (Aebi and Richterich, 1963; Markert and Møller, 1959). The number of enzymes consisting of several isoenzymes is growing rapidly. They are all composed of subunits, i.e. they contain more than one structurally distinct polypeptide chain, each being determined by a separate gene locus. Most often the active enzyme is formed of two types of subunits. If it is a tetramer (e.g. lactate dehydrogenase),

5 different isoenzymes, 2 co-polymers (A4, B4) and 3 hybridpolymers (AB_3, A_2B_2, A_3B) exist [Richterich and Burger, 1963 (1); Richterich, Schafroth and Aebi, 1963 (2)]; if it is a dimer (e.g. creatine phosphokinase), 3 isoenzymes are formed (Dawson, Eppenberger and Kaplan, 1965; Eppenberger, Eppenberger, Richterich and Aebi, 1964; Pette and Bücher, 1963). Due to differences in excess charge of the molecules they can easily be separated by electrophoresis.

Minor differences in enzyme structure, and consequently in physicochemical properties, may also exist from one individual to another. Thus, what is known for the blood group substances is also true for many enzymes. These genetically determined polymorphisms, as they are observed in many enzymes, offer evidence that genetic diversity in a human population is reflected to a considerable extent in enzyme diversity. Such differences in the enzyme level or (iso)enzyme pattern between individuals may be due to either variation in the qualitative characteristics of the enzyme they synthesise or to differences in rates of synthesis or degradation. The study of biochemical individuality has revealed that there are no basic differences between enzyme polymorphisms, enzyme variants and enzyme deficiency conditions, either silent or leading to an "inborn error of metabolism" (Aebi, 1967; Harris, 1966). The arbitrary distinction may be made according to the activity level of the mutant enzyme and to the frequency of the gene responsible for this mutation.

Red cell acid phosphatase is a typical example for enzyme polymorphism. As shown by Harris et al. (1968), six different phenotypes can be distinguished by means of the isoenzyme pattern obtained by electrophoresis (Fig. 8).

Three allelic genes (P^a, P^b, P^c) at an autosomal locus are involved. They are not only responsible for the characteristic isoenzyme pattern, but also contribute to a different extent to the level of total enzyme activity ($P^a:P^b:P^c \sim 2:3:4$). Therefore, if total red cell acid phosphatase activity is determined in a series of randomly selected individuals, a broad continuous unimodal distribution curve is obtained. However, if this is done separately for each of the phenotypes, the shape of the resulting curves indicates a considerably smaller degree of variance (Fig. 9).

Therefore the range of normal variation can be narrowed considerably if an individual value can be attributed to a certain phenotype. This is of practical importance whenever enzyme activity data obtained in survey studies of any kind have to be evaluated. It is very likely that this type of polymorphism is not an exception but represents a rather normal situation in enzymology. In fact, among 10 arbitrarily chosen enzymes, Harris (1967) and his collaborators (1968) have detected three quite striking examples of genetically determined polymorphisms. Recently this list has even been extended. Based on the data shown in Table 1, the probability of two randomly selected individuals being the same phenotype has been calculated. No doubt the preliminary figure of 0.007

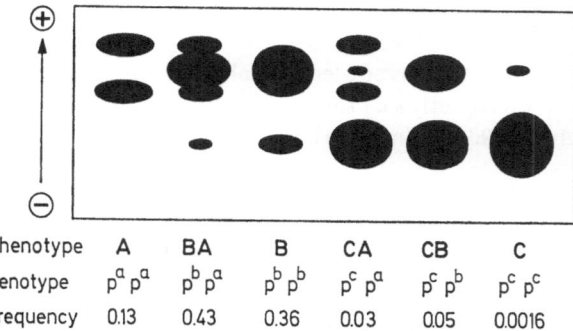

Phenotype	A	BA	B	CA	CB	C
Genotype	$p^a p^a$	$p^b p^a$	$p^b p^b$	$p^c p^a$	$p^c p^b$	$p^c p^c$
Frequency	0.13	0.43	0.36	0.03	0.05	0.0016

Fig. 8. Diagram of isoenzyme components seen in the various red cell acid phosphatase phenotypes after electrophoresis at pH 6.0. (Harris, 1968)

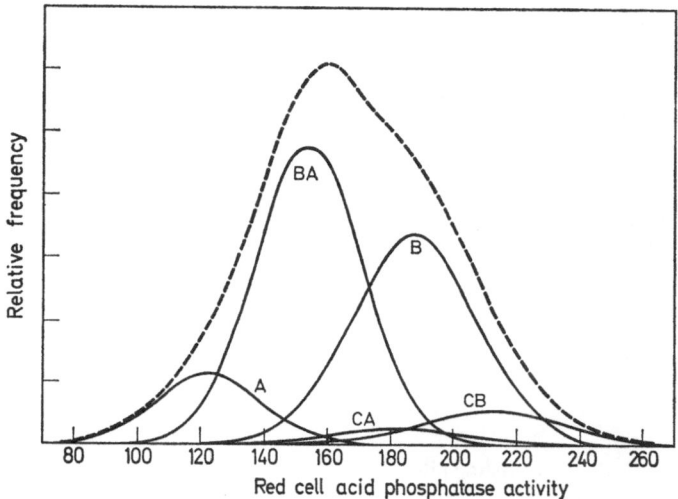

Fig. 9. Distribution of red cell phosphatase activities in the general population (top line) and in the separate phenotypes. Phosphatase activity expressed in μ Moles p-nitrophenol liberated in 30 min at 37 °C per g hemoglobin. (Harris, 1968)

(= 1:140) based on nine such cases is still much too high (Harris et al., 1968).

A similar situation is found in rare anomalies such as enzyme deficiency due to the synthesis of a less active or an unstable enzyme variant. Often even homozygotes do not show any clinical symptoms, but exert abnormal response only after provocation, such as intolerance to drugs (e.g. primaquine in glucose-6-phosphate dehydrogenase deficiency) or certain components of normal food (e.g. fructose in fructose intolerance or galactose in galactosemia) (Aebi, 1967). The opposite situation in mutants with an abnormally high enzyme activity may be found, too. An interesting example is the atypical liver alcohol dehydrogenase, an enzyme variant found in one out of 7 to 10 individuals (Table 2). At pH 7 the liver alcohol dehydrogenase of these humans is about five times as active as the normal enzyme (von Wartburg, Papenberg and Aebi, 1965; von Wartburg and Schürch, 1968) (Fig. 10). It is

Table 1. *Enzyme polymorphism in the English population*
(Harris, 1967)

Enzyme	Number of alleles with frequency greater than 0.01	Frequency of commonest phenotype	Probability of two randomly selected individuals being of the same phenotype
Red-cell acid phosphatase	3	0.43	0.34
Phosphoglucomutase			
Locus PGM_1	2	0.58	0.47
Locus PGM_3	2	0.53	0.50
Placental alkaline phosphatase	3	0.41	0.31
Acetyl transferase	2	0.50	0.50
Adenylate kinase	2	0.90	0.82
Serum cholinesterase			
Locus E_1	2	0.96	0.92
Locus E_2	2	0.90	0.82
6-phosphogluconate dehydrogenase	2	0.96	0.92
Combined		0.02	0.007

probable that many other silent anomalies of this type, not yet detected, do in fact exist.

What has this statement to do with nutrition, or even with malnutrition? We still do not know exactly why, in a population or in a family, some individuals show symptoms of malnutrition whereas others do not. In this respect studies on "pair fed" children are particularly striking. Why does one of them thrive quite readily and the other fail to do so? Is this exclusively due to minor qualitative or quantitative differences in food intake or might it be that in a marginal situation genetic factors such as enzyme polymorphisms or enzyme variants play a role? It will be an interesting and promising task to study known and search for new

Table 2. *Properties of normal and atypical human liver alcohol dehydrogenase* (von Wartburg et al., 1968)

Property	Normal	Atypical
Specific activity (IU/mg prot.)	ca. 3	ca. 10
Total activity (IU/liver)	2'700	16'200
pH optimum (ethanol)	10.8	8.5
I_{50} with OP (M)[a]	6.7×10^{-5}	3.3×10^{-4}
Effect of 0.66 M thiourea (V_{TH}/V_C)	2.20	0.65
Substrate specificity		
1.6×10^{-2} M Ethanol	100%	100%
Butanol	87%	54%
3.3×10^{-4} M Benzyl alcohol	85%	22%
Cyclo-hexanol	82%	16%
K_M for ETOH, ACALD, NAD, NADH$_2$	No detectable difference	
Molecular weight (Sephadex G-200)	No detectable difference (87'000)	

[a] I_{50} = 50% inhibition in presence of ortho-phenanthrolin

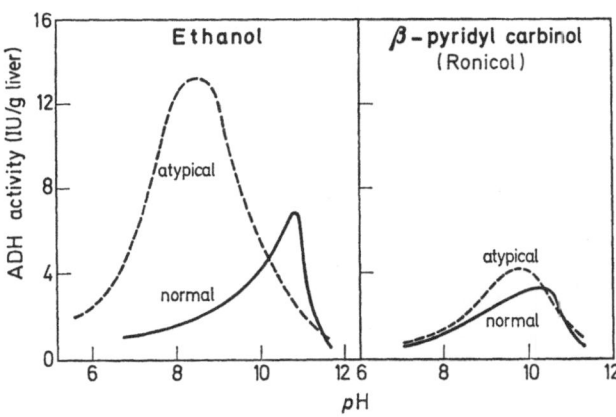

Fig. 10. pH-rate profile of alcohol dehydrogenase of human liver. — normal enzyme; - - - atypical form. Left: substrate ethanol; Right: substrate β-pyridyl carbinol. (von Wartburg et al., 1968)

polymorphisms in enzymes relevant for the utilisation and the metabolism of proteins, amino acids and other components of man's nutrition.

Technical Aspects of the Study of Enzymes and Enzyme Patterns in Humans

Finally, some methodical possibilities for the investigation of correla-
tions between enzyme levels and nutritional status in man will be discussed. There is a variety of approaches, direct and indirect, each of them
having its own advantages and drawbacks. As demonstrated by recent
progress in this field, the following possibilities have to be considered:

1. Enzymes in Tissues

Analysis of enzymes in tissues: Samples of liver and muscle can be
obtained by needle biopsy, intestinal mucosa by the suction biopsy technique. The quantity of test material available (5 to 15 mg) permits the
simultaneous analysis of various enzymes provided that suitable ultra-
micromethods are used. Some data on enzyme activity in liver of cases of
kwashiorkor have been summarised some time ago (Aebi, 1964; Waterlow, Cravioto and Stephen, 1960). More recent studies in malnourished
children, based on the needle biopsy technique, have shown that this
approach may provide most informative results. Stephen and Waterlow
(1968) have measured the activity of enzymes related to protein
metabolism in liver biopsies from 22 malnourished and recovering

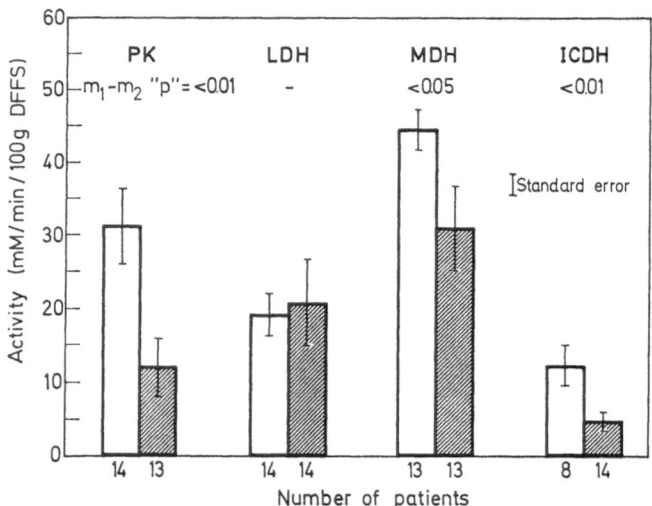

Fig. 11. Enzyme activities in muscle biopsy specimens of malnourished (■) and
normal children (□). PK pyruvate kinase; LDH lactate dehydrogenase; MDH
malic dehydrogenase; ICDH isocitrate dehydrogenase. (Metcoff, 1968)

Jamaican infants. Amino-acid activating enzyme levels were increased initially and fell on recovery; however, argininosuccinase activity was low initially and rose in recovery. Metcoff (1968), on the other hand, used muscle biopsy specimens for a study of ionic composition and cell metabolism with protein-calorie malnutrition in man. From his data on enzyme levels shown in Fig. 11, it can be seen that there is a decrease in all instances with the exception of lactate dehydrogenase. The decrease is most pronounced in pyruvate kinase and isocitrate dehydrogenase activity.

However, it must be stressed that the main obstacles in performing studies along this line are not of technical, but rather of psychological nature. Practical considerations make it often impossible to analyse tissue samples which can be obtained by biopsy only. Therefore, various efforts have been made to include the analysis of skin or mucosa specimens, or even hair or nail clippings in the study of malnutrition.

2. Analysis of Enzymes in Blood

The analysis of enzyme levels in serum, which is very common in clinical chemistry, is another promising field for the evaluation of the nutritional status. A considerable number of serum enzymes are reduced by a factor of 2 to 4 in protein-calorie malnutrition, such as cholinesterase, esterase, amylase and alkaline phosphatase (Aebi, 1964; Waterlow, Cravioto and Stephen, 1960). Presumably those enzymes may be most sensitive reagents which show an extreme partition between tissue and serum. In creatine phosphokinase e.g. the ratio between the normal activity level in muscle tissue and in serum is about $10^6:1$ [Richterich, Rosin, Aebi and Rossi, 1963 (3)]. However, there are many factors affecting this ratio, such as muscle function (Wiesmann, Moser, Richterich and Rossi, 1965), genetic factors [Richterich, Rosin, Aebi and Rossi, 1963 (3)] and vitamin-E deficiency (Olson, 1967). Particularly when an organ biopsy is not possible, the analysis of erythrocyte and leucocyte enzymes or isoenzymes may serve as a valuable substitute (Andersen, Gerhard and Clausen, 1964). In the former, deviations from normal are relatively small, whereas in the latter, relatively large quantities of starting material are required.

3. Enzymes in Urine

More than 30 enzymes are known to be excreted in urine (Dubach, 1968). Despite many difficulties, (e.g. due to the presence of inhibitors, denaturating action of urea, bacterial contamination) the easy accessibility of the material may perhaps also contribute to another approach in the study of malnutrition.

3*

Several recent investigations have led to results which favor this hypothesis. Ittyerah et al. (Ittyerah, Dumm, Bahhawat, 1967) have shown that arylsulfatase A excretion in urine[1] is significantly higher in children with kwashiorkor (Fig. 12).

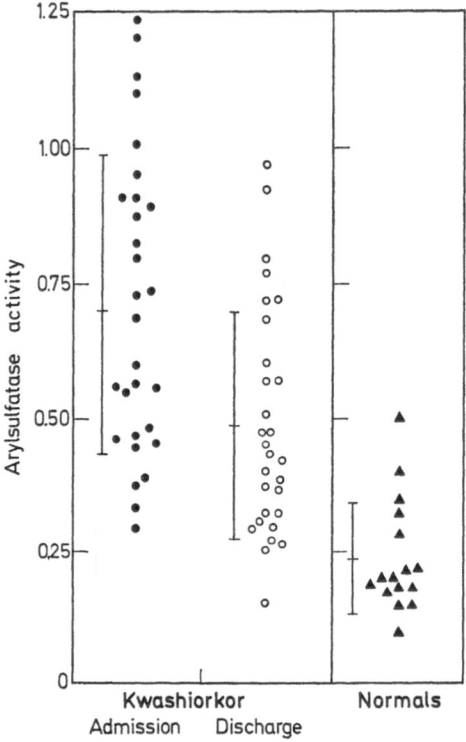

Fig. 12. Arylsulfatase A activity of random urine specimens from children with kwashiorkor on admission to hospital and on discharge, and of normal controls. The mean ± 1 standard deviation is shown in the figure. (Ittyerah et al., 1967)

Since other lysosomal enzymes, such as arylsulfatase B and acid phosphatase, do not show any change, this phenomenon has to be considered as rather specific. However, this seems to be a complex phenomenon, since in vitamin-A deficiency a similar alteration can be observed.

The aim of this report has been to discuss some relations between nutrition and enzyme levels or enzyme patterns in tissues and body fluids. However, hormonal, genetic and circadian influences have been included as well, in order to demonstrate how many obstacles and pitfalls

---

[1] The arylsulfatase A catalyses the following reaction:

p-Nitrocatecholsulfate $\longrightarrow$ p-Nitrocatechol + SO_4^{--}.

may exist. Despite all the difficulties which have to be faced when enzymes are studied in subclinical, marginal and severe cases of malnutrition, our efforts to get more insight by finding better indicators should be increased. It is not unreasonable to assume that further research in clinical enzymology will help us in making a more precise diagnosis of the different stages and the various forms of malnutrition.

From the University of Cambridge and the Medical Research Council, Dunn
Nutrition Laboratories, Cambridge (England) and Child Nutrition Research Unit,
Box 7051 (MRC), Kampala (Uganda)

Factors Which May Affect the Biochemical
Response to Protein-Calorie Malnutrition

R. G. Whitehead

With 1 Figure

Contents

Introduction

At a recent meeting of nutritional scientists in England, Dr. Cicely
Williams made the following very important comment. She described
how, until 1950, she had pleaded with the medical profession to recognise
that kwashiorkor was caused by protein deficiency. Since 1950, however,
she has been insisting that kwashiorkor is not just protein deficiency but
many other things as well. I am sure that this warning to the scientists
was a timely one; pediatricians who are trying to treat and cure malnourished children are only too aware of the complexity of protein-
calorie malnutrition, but there is an unfortunate gulf between their experiences and the investigations of the medical scientists. There has been
a tendency to equate kwashiorkor metabolically solely with protein
malnutrition, and to consider marasmus as purely calorie undernutrition. The more general term protein-calorie malnutrition has been considered in equally elementary terms. This over-simplification is dangerous
if our work is to have any significance in terms of practical nutrition and
the relief of human suffering. I fear this over-simplification could be the
cause of controversy which might cloud the prospects of international
cooperation in a field where it is of paramount importance. This topic is

not a new one, it has been discussed before, but it is of such basic importance that I believe it would be profitable to consider it once again. In this paper I will consider some of the factors which may modify the biochemical response to malnutrition.

1. Relative Proportions of Proteins and "Calories" in the Diet

The first variable is well known. My own practical experience was gained in the district around Kampala. Here the dietary etiology of kwashiorkor is relatively simple. The habitual food is matooke, or steamed plantain; it has a very low-protein, but a high-carbohydrate content. Thus, in kwashiorkor we have a fairly pure form of protein malnutrition. Primary calorie insufficiency is relatively rare, and usually results from failure of lactation. Almost certainly it is for this reason that

Table 1. *Biochemical abnormalities found in two extreme forms of severe protein-calorie malnutrition*

	Low-protein/ high-calorie diet	Low-protein/ low-calorie diet
Serum aminoacid pattern	Distorted	Normal
Serum albumin	Low	Normal
Imidazoleacrylic acid excretion	High	None
Phenylalanine/tyrosine ratio	High	Normal
Skin fat content	Normal or high	Low
Liver fat	High	Low

we have been able to describe such marked biochemical differences between these two forms of protein-calorie malnutrition in Uganda. Similar differences have been reproduced in animal experiments designed to imitate these two extreme ends of the protein-calorie malnutrition spectrum. Table 1 lists some of my own findings.

Such clear-cut differences, however, are not found in children or animals in whom the dietary etiology has been more complicated. For example, when a low-protein diet has resulted in a prolonged lack of appetite, or when a child or animal has become so ill that complete anorexia has resulted. Diarrhoea or vomiting can also affect the usual biochemical pattern by severely limiting the availability of total calories. In other countries, in particular where people have moved to live in cities, the protein-calorie ratio can be very variable; children have to eat what is available. It is little wonder that the biochemical patterns in such children are difficult to interpret. We need to study these complexities in more detail than has previously been the custom. This is equally important to those of us who carry out planned nutritional experiments in animals and to those who investigate human malnutrition.

2. The Chemical Nature of the Dietary Proteins and "Calories"

The second factor which I suggest might vary the metabolic response to protein-calorie malnutrition is the chemical nature of the proteins and "calories" in the diet. The amino-acid composition of the dietary proteins is a well-known variable, it is of course the biological value of the protein rather than "N × 6.25" which is important. The chemical nature of the "calories" has not received so much attention, but animal experiments have clearly demonstrated that if the diet is imbalanced with fat rather than with carbohydrate the same metabolic abnormalities are not found. Heard et al. (1968) reported, for example, that the serum amino-acid pattern fails to become distorted when excess fat rather than excess carbohydrate is fed. Bender (1968) has suggested that the abnormalities in fat metabolism which accompany kwashiorkor may differ if the main source of carbohydrate is sucrose rather than starch. If this proves to be the case, it will be most important. In Uganda, as in most other countries, the carbohydrate involved is usually starch, but occasionally, in young babies, kwashiorkor results from feeding watered-down milk containing generous amounts of sucrose to make it more palatable. Similarly in the Sudan, some babies are fed on sweet dilute fruit juice. Sucrose is often the carbohydrate involved in the West Indies. I have not yet investigated the relative effects of sucrose and starch on a low-protein diet, but nevertheless I am sure we should not talk too glibly just about "calories".

3. Subsidiary Dietary Deficiencies

The introduction of the term protein-calorie malnutrition was valuable because it provided a way of describing the wide range of clinical abnormalities found in various malnourished children, but its uncritical use risks ignoring other dietary deficiencies which might also be present. This is an important consideration since they can vary widely from one area to another. These super-imposed deficiencies take the form of various vitamins and minerals, many of which are important co-factors in enzyme reactions, and it is always possible that they may distort the metabolic picture found in less complicated malnutrition.

In Uganda, we encountered this problem when we studied histidine metabolism in severe kwashiorkor (Table 2) (Whitehead, 1964). After a test dose of histidine, imidazoleacrylic acid was excreted in severely ill children because of the deficiency of the enzyme "urocanase", but after treatment on a high-protein diet the metabolism of histidine was corrected, and imidazoleacrylic acid no longer appeared. In a few children, however, formiminoglutamic acid, which previously had not been present in the urine of the untreated child, was now detectable in significant amounts. The reason for this became apparent when it was shown that a small percentage of the children exhibited a megaloblastosis as well as kwashiorkor. The formiminoglutamic acid did not disappear until extra folic acid was added to the diet. Had this difference in metabolic response been found in different laboratories rather than the same one, you can imagine what controversy could have developed.

Table 2. *Histidine metabolism in kwashiorkor*

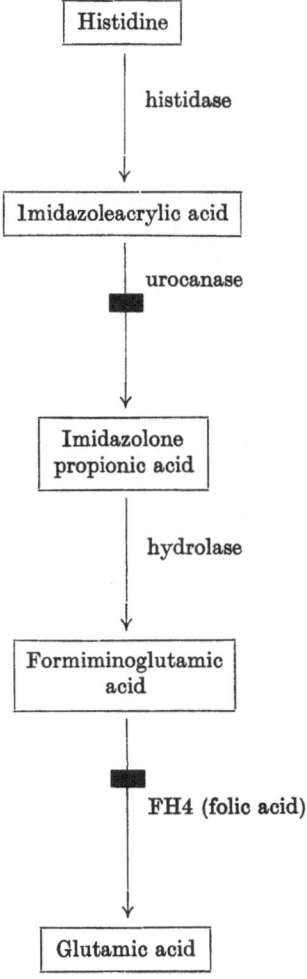

Another deficiency which seems to be important in Uganda is that of magnesium. Potassium deficiency has also been suspected, but it does not seem to be the problem it is in the West Indies. This is not really surprising since matooke contains much potassium. Vis, working in the eastern Congo, has suggested that many of the abnormalities in mineral metabolism described in kwashiorkor are not primarily due to the pathological effects of protein lack but to supplementary deficiencies of the various minerals in the diet (Vis, Dubois, Vanderborght and DeMaeyer, 1965).

Kwashiorkor complicated by a deficiency in vitamin A is reported from many parts of the world, especially Central America. Since vitamin A may be essential for the structural integrity of various subcellular components, it could have an important bearing on the cell's response to protein-calorie malnutrition. In other countries, especially in desert or semi-desert areas, vitamin C can also be deficient, and in Egypt and South Africa vitamin D may also be lacking. The latter deficiency produces the interesting phenomenon of rickets in the presence of normal or even reduced serum alkaline phosphatase levels.

Secondary dietary deficiencies should be anticipated in substandard foods and their possible importance must be considered when the results of one laboratory are being evaluated in another country.

4. Genetic Factors

It would be surprising if there were no genetic factors involved in the metabolic response to malnutrition. Various individuals and peoples vary in their susceptibilities to other stresses and there is no reason why nutritional stress should be an exception. A specific example of the intervention of a genetic factor is indicated by our investigations on jejunal lactase levels. The work was initiated because of a suspected lactose intolerance in certain malnourished children (Cook, Lakin and Whitehead, 1967). This is an interesting problem since the different ability of various malnourished children to digest lactose is the cause of divided opinion on the value of dried skimmed milk as a therapeutic diet. One explanation for this controversy could be that different people inherit different levels of jejunal lactase. In those who normally have relatively low levels the effect of malnutrition may result in the enzyme activity falling below a critical level, but in others the "functional reserve" of the organ as a whole may allow lactose digestion to continue practically normally. In Uganda, the Bantu tribes appear to inherit much lower jejunal lactase levels than their Nilotic compatriots (Cook and Kajubi, 1966). Even in well-nourished Bantu people, for example Baganda doctors, a test dose of lactose can produce diarrhoea. Clearly in such people, any factor which lowered lactase levels, even slightly, might have quite a profound effect. Hansen and his colleagues working in Cape Town have emphasised the problem of lactose intolerance in *their* malnourished Bantu children.

It would be possible to give other specific examples of congenital abnormalities which could predispose a child to nutritional stress, but I am sure there must be many, *more subtle*, variations in the individual's metabolic make-up, which are of more general importance. It is these slight differences in metabolic patterns which make one individual different from another and which give the various peoples of the world their characteristic appearance. I am sure we should give the possibility of genetic variations careful consideration when we try to correlate nutritional findings from different countries. Dr. Aebi has already discussed this matter when he spoke about biochemical individuality. I would like to join with him in emphasising its importance.

5. Age

It is commonly stated that kwashiorkor is a problem encountered in children aged about 2 years. Whilst in general this is true of Uganda, it is not so in other parts of the world. For example, in Sudan and Jamaica, the peak incidence is about 9 or 10 months, whilst in Guatemala I was shown many cases that were about 4 to 5 years old. In Uganda I have occasionally seen children aged 10 to 12 years with kwashiorkor. Nutritional marasmus can, of course, occur at any age, although my own studies in Uganda have been mainly confined to babies aged about 6 months.

But are we right to assume that the metabolic response to nutritional stress should be the same regardless of age? Surely not. Cell structure, enzymic patterns and homeostatic balance change progressively with age and it is upon these metabolic variations that the biochemical stresses of malnutrition are superimposed. It is clearly possible that different aspects of metabolism may be more susceptible at different ages. Our own experiments in animals in Cambridge have illustrated this complication. In Uganda, kwashiorkor is rare in children over 4 years and uncommon in children older than 3 years, although there is no reason to believe that diet improves significantly with increasing age. Presumably the metabolic needs imposed by a more rapid rate of growth and muscle synthesis make the younger children more sensitive to protein deprivation. In Table 1, I showed differences between my findings in kwashiorkor and marasmus. The kwashiorkor cases were about 2 years old, whilst the ones with marasmus were only 6 months. It might be well worthwhile reconsidering how much these differences were really due to the diet, and how much to age!

6. The Period of Chronic Malnutrition before the Acute Episode of "Clinical" Malnutrition

In connection with age it is natural to consider the importance of the period of chronic malnutrition prior to the acute episode of kwashiorkor or marasmus. This implies acceptance of the hypothesis that a child is gradually weakened by a prolonged phase of malnutrition but the onset of clinical kwashiorkor develops suddenly, possibly as a result of some other stress like an infection.

It is almost impossible to assess the length or severity of the malnutrition during this chronic episode, since accurate dietary histories are difficult to compile from information supplied by the mother. I tried to get round this difficulty by sub-classifying the children with kwashiorkor according to the condition of their hair (Whitehead, 1967). In Uganda,

Table 3. *Percentage incidence of malaria, sickling, and worm infestations in children admitted for treatment of kwashiorkor*

	Pale-haired	Dark-haired
Malaria	14	36
Roundworm	0	21
Sickling test positive	9	7
Homozygous sickle cell anemia	0	0
Severe hookworm	18	50
Mean egg count in infested cases (per g feces)	3670	14500

Table 4. *Biochemical and anthropometric measurements in children admitted for treatment of kwashiorkor*

	Pale-haired mean	Dark-haired mean	Normal mean
Weight for age (% of normal)	72	71	100
Triceps skinfold (mm)	7.6	6.5	10.0
Serum protein (g/100 ml)	4.7	4.2	7.0
Serum amino-acid ratio	5.7	5.7	1.8
Blood glucose (mg/100 ml)	47.0	58.0	68.0
Blood Hb (g/100 ml)	8.2	6.7	12.0
Urinary hydroxyproline index	0.94	1.7	2.8

well-fed children have a thick curly mop of black hair, but prolonged feeding on a low-protein diet eventually results in pale, sparse hair. I assumed that, in cases of kwashiorkor in which the hair was not so discolored, the malnutrition had not been as prolonged as in those children with much paler hair. In Table 3, I have listed the parasite loads in the two types of children. The higher incidence of infestations in the darker-haired children does indicate that the onset of kwashiorkor might have been of a more acute nature in these children.

In Table 4 are summarised the biochemical findings in the two types of children. Many features are different. Of particular interest are the blood glucose levels. There is much controversy on whether hypoglycemia is, or is not, a problem in kwashiorkor. For example, in Johannesburg, Wayburne has told me that it is the most common cause of death whilst in Cape Town, Hansen said hypoglycemia is rarely encountered. In our ward in Kampala we obviously have both situations, and we may have the rational explanation. The low serum protein levels in the darker haired children could be due more to parasite infestation than to dietary deprivation of protein. Another feature of Table 4 is the higher hydroxyproline excretion; such values are often found in these complicated cases.

These results do demonstrate the importance of this period of chronic malnutrition on the ultimate nature of the biochemistry in kwashiorkor. This chronic phase must vary widely in different environments and more consideration of its importance may help to rationalise some of the anomalous results obtained from different laboratories.

7. Infection and Worm Infestations

Much has been written about the importance of infections in malnourished children but this has mainly been in a clinical context. This inseparable relationship is equally important to nutritional scientists, especially to those like myself who are trying to develop biochemical tests for the evaluation of nutritional status. I have already shown how, in children with severe worm infestations, the low excretion of hydroxyproline peptides normally found in malnourished children does not occur; normal or even elevated levels are present.

The experience of Hansen in South Africa with serum amino-acid patterns is another example of this problem (Truswell, Wannenburg, Wittman and Hansen, 1966). In Cape Town, kwashiorkor is invariably preceded by infective diarrhoea and vomiting. Elevated amino-acid ratios of the type described in Uganda are not found, and it is postulated that the severe calorie deprivation which is superimposed on the primary protein malnutrition reverts the serum amino-acid pattern to one more characteristic of marasmus.

These two examples both illustrate cases where infections or infestations cancel out the abnormal metabolic pattern found in more simple forms of malnutrition. On other occasions, the biochemical response to infection may be in the same direction as to the dietary deficiency. An example of this is severe hookworm infestation. Both protein malnutrition and this worm infestation eventually lower the serum albumin level but, if this criterion were being used as an index of malnutrition, an erroneously low value would be obtained relative to the degree of dietary stress.

A dramatic example of how infections can affect the biochemistry of malnourished children is shown in Fig. 1, which illustrates hydroxyproline excretion, expressed as the hydroxyproline index, in a child recovering from nutritional marasmus. Each episode of infection markedly affected excretion and these correlated with minor fluctuations in weight.

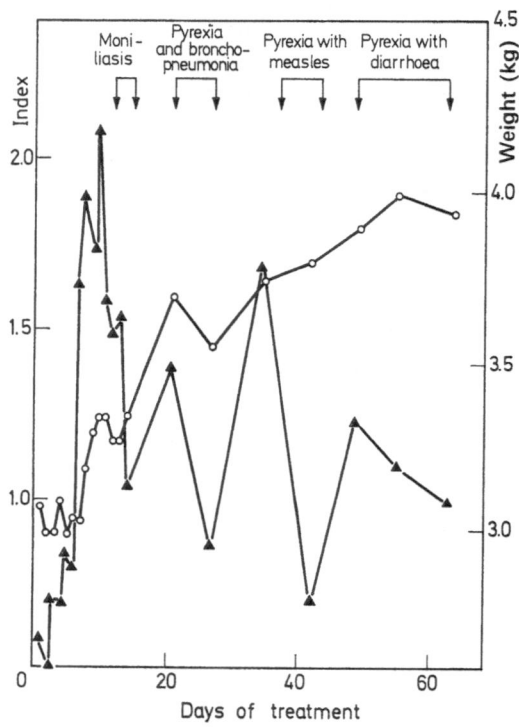

Fig. 1. Changes in the urinary hydroxyproline index and the body weight during the treatment of a child with nutritional marasmus. The incidence of infection is also shown. ▲ — ▲ Hydroxyproline index; o — o body weight

Conclusion

At the beginning of this paper I said I would not be providing new information. My aim was to illustrate the complexity of the problem which faces biochemists. Our pediatric colleagues refer to kwashiorkor, marasmus and protein-calorie malnutrition as syndromes, not single disease states. This is all too true. Under normal circumstances, biochemists would consider a problem with so many independent variables an impossible one for scientific study, but the humanitarian need for information prevents us from resigning our task. The enormity of the problem makes it unlikely that any single scientist in any single country will provide a complete solution, and international cooperation is probably more important in this than in any other field of science. If this is to be successful, our individual results must be correlated with due regard to our very differing environments. We should try to be objective in discussing our results, we should avoid allowing them to become the source of unconstructive controversy.

From the Institute of Nutrition of Central America and Panama (INCAP),
Guatemala, C. A.

Proposed Methodology for the Biochemical Evaluation of Protein Malnutrition in Children*

G. Arroyave

With 5 Figures

Contents

Introduction

The existence of cases of kwashiorkor in a population group is the most dramatic expression of protein deficiency, but the underlying mild to moderate protein malnutrition affecting the vast majority of children in population groups of low socio-economic level constitutes the problem of major magnitude from the public health point of view. The assessment of the extent to which this nutritional problem is affecting population groups is basic to its solution.

The present paper proposes that four relatively simple biochemical methods, when applied together, can give valuable information regarding the state of protein nutrition of children. Fig. 1 presents, in a simplified manner, the main stages of protein metabolism and the corresponding methods which will provide information regarding the level of efficiency at which they are operating. A discussion of these methods is presented in the following paragraphs.

1. Plasma Free Amino Acids

Alterations in the pattern of free plasma amino acids have been demonstrated under conditions of a restricted protein intake (Arroyave, 1963). A few days on a protein-free diet brings about, in children, a decrease in many essential amino acids and an increase in some non-

* INCAP Publication I-471.

essential ones. The non-essential to essential amino-acid ratio is, therefore, elevated. These changes are illustrated in Fig. 2 and, although of lesser magnitude, resemble those observed in naturally occurring severe protein malnutrition of the kwashiorkor type. The pattern observed in the latter situation is shown in Fig. 3 and is consistently found in our studies at INCAP (Arroyave et al., 1962) and by other workers (Westall et al., 1958; Edozien et al., 1960; Cravioto et al., 1960; Norton, 1960; Saunders et al., 1967). On the basis of these detailed amino-acid studies, a simple paper chromatographic technique has been devised by Whitehead (1964) to determine the ratio of some essential amino acids, prin-

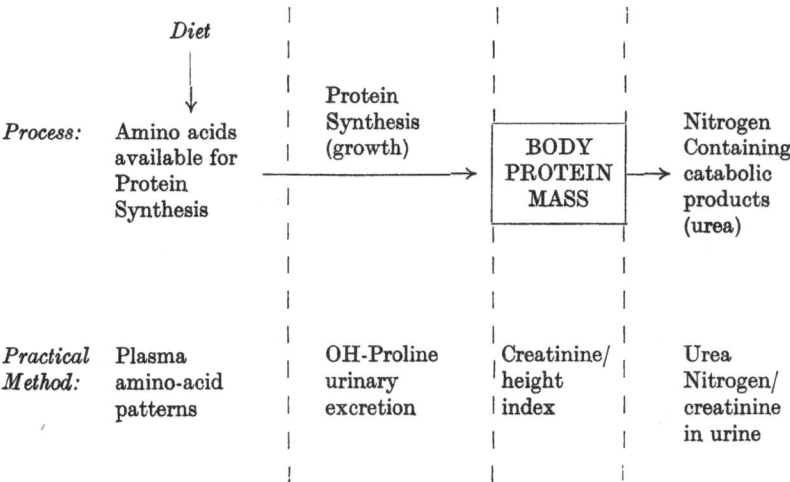

Fig. 1. Proposed methodology for the biochemical evaluation of protein malnutrition in children

cipally valine, leucine and isoleucine, to a group of non-essential amino acids. The application of this simple method has also been confirmatory of the findings described previously (Rutishauser and Whitehead, 1969; Arroyave and Bowering, 1968). A high non-essential to essential ratio is found under conditions of low protein intake and in severe protein malnutrition. Treatment results in a return to a low normal value.

Some exceptions to this conclusion have been recently presented (McLaren et al., 1965; Ittyerah et al., 1965; Truswell et al., 1966; Amasuya and Narasinga Rao, 1968), but the inconsistencies should not be considered sufficient yet to invalidate the method proposed. More research is needed to determine the reasons for such inconsistency. The result of this research would contribute to the knowledge of the factors which produce the changes in amino-acid pattern, and of the circumstances which modify the response, adding consequently to the value of

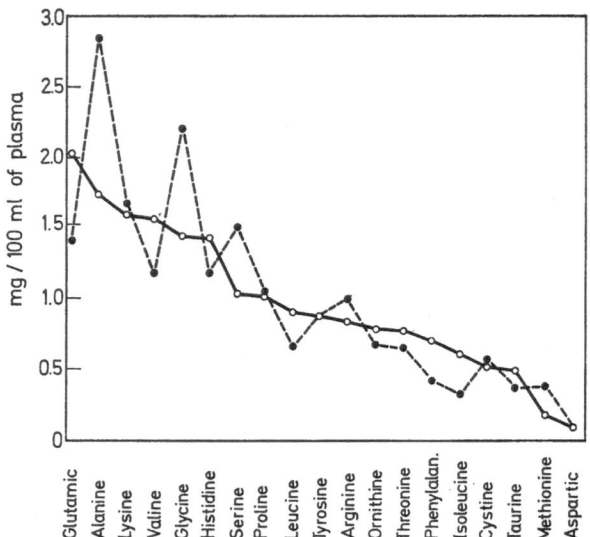

Fig. 2. Effect of a nitrogen-free diet on the plasma amino-acid pattern. Child S.A. (90 cal/kg/day). ---- Pattern after 7 days on N-free diet; —— Pattern of control well nourished children. Taken from: Arroyave, G.: Biochemical signs of mild-moderate forms of protein-calorie malnutrition. In: Blix, G., ed. *Mild-moderate forms of protein-calorie malnutrition*. Symposia of the Swedish Nutrition Foundation I, Bastad, August 29—31, 1962. Uppsala: Almqvist & Wiksells 1963, p. 32—46

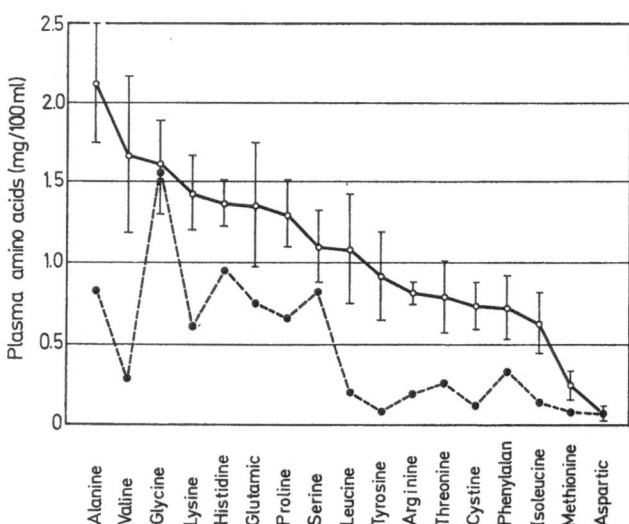

Fig. 3. Fasting plasma amino acids in six children with kwashiorkor. ●----● Kwashiorkor (average of six cases); o——o recovered children (average of five cases). Taken from: Arroyave, G.: Biochemical signs of mild-moderate forms of protein-calorie malnutrition. In: Blix, G., ed. *Mild-moderate forms of protein-calorie malnutrition*. Symposia of the Swedish Nutrition Foundation I, Bastad, August 29—31, 1962. Uppsala: Almqvist & Wiksells 1963, p. 32—46

this approach. At the present time, it is proposed that this simple technique should be applied as a way to detect alterations in the free amino acid pool consequent to an insufficient supply of dietary protein in relation to the requirements. Direct studies in children of the level of protein intake per kilogram of body weight necessary to maintain a low normal ratio are being conducted in our laboratories.

2. Hydroxyproline Excretion

The excretion of the amino acid hydroxyproline in the urine has been proposed as an indicator of the rate of growth of children. Several workers have shown that it is markedly reduced in children with kwashiorkor (Picou et al., 1965; Whitehead, 1965; Howells et al. 1967; Wharton, et al., 1967). It also seems to correlate with the percent deficit of standard weight in children from population groups of low socio-economic level, although such correlation may not necessarily be true under all circumstances. We must remember that the hydroxyproline excretion gives the "present" situation, while the weight deficit is a "history". At the moment a child or group of children are studied, they may be undergoing a nutritional situation which limits the rate of growth, while during their previous life they may not have suffered from any limitation in their food intake and therefore their body weight may not be significantly affected. The converse could be also true. We must conclude, however, that a low hydroxyproline excretion is indicative of a situation of slow growth rate at the time of the examination. As an example of actual data in relation

Table 1. *Urinary hydroxyproline index in Ugandan children*

Children's groups	No.	Hydroxyproline index (mean $\pm$ S.D.)
Professional class		
Aged under 4 years	42	2.9 $\pm$ 0.9
Aged 4—7 years	16	3.0 $\pm$ 0.8
Kwashiorkor patients	10	0.9 $\pm$ 0.3
Marasmic patients	8	1.1 $\pm$ 0.4
Clinic children		
Below 70% standard weight	6[a]	1.1 $\pm$ 0.4
71—80% standard weight	9	1.4 $\pm$ 0.5
71—80% standard weight	21[a]	1.9 $\pm$ 0.8
81—90% standard weight	15	1.7 $\pm$ 0.6
81—90% standard weight	33[a]	2.1 $\pm$ 1.0
Above 90% standard weight	26	2.7 $\pm$ 1.0
90% standard weight	22[a]	2.2 $\pm$ 0.7

[a] Receiving supplementary milk packets.
Taken from: Whitehead, R. G.: Lancet **1965 II,** 567.

to this index, results from the work of Whitehead (1965) are reproduced
here (Table 1 on p. 51).

3. Creatinine-Height Index

Ultimately, a situation of insufficient protein intake results in a
decrease in the protein mass of the body. Therefore, the detection of this
decrease would be a direct indication of protein depletion. The excretion
of urinary creatinine per unit of time is recognised as an indirect indicator
of muscle mass. Recently a creatinine-height index has been proposed as
a simple quantitative expression of the extent of this depletion (Viteri,
Arroyave and Behar, 1966). The index is the ratio of the amount of
creatinine excreted per unit of time by the children under study to that
excreted by normal children of the same height regardless of age. A ratio

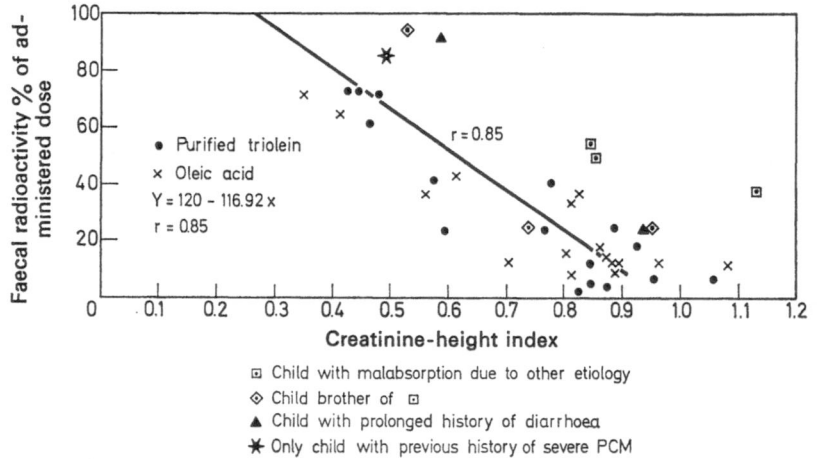

Fig. 4. Correlation of fecal radioactivity in percent of administered dose of I[131] labelled fat and creatinine-height index. Data kindly supplied by Dr. Fernando Viteri at INCAP

of one would of course indicate a fully developed muscle mass for the
subjects being assessed. Ratios lower than 0.9 are interpreted as indicative of proportionally depleted protein mass. Results obtained in our
laboratories have demonstrated the value of this index and its significance
in terms of function. Children recovering from severe protein malnutrition do not regain, for example, normal intestinal absorption until their
creatinine-height index has returned to a ratio near one (Viteri). This is
illustrated in Fig. 4. In another study, it was shown that the index was
significantly reduced in a group of children of low socio-economic level,
after they had suffered from an epidemic of measles, indicating the impact
of this infectious stress on the protein nutritional status of those children
who were subsisting on an already marginal or low protein intake (Viteri).

4. Urinary Excretion of Urea Nitrogen

Children suffering from dietary protein deficiency have a low excretion of urea nitrogen as determined in a fasting urine sample (Arroyave, 1963). The ratio of urea nitrogen to creatinine is useful since it does not require the collection of a timed urine specimen. It must be emphasised that a low urea nitrogen to creatinine ratio does not necessarily indicate that children are suffering from protein-calorie malnutrition, but only that at the moment of the examination they were subsisting on a low

Table 2. *Relative excretion of urea, creatinine, and total nitrogen in Central American children under different nutritional conditions, 1961*

Determinations in fasting urine sample		Low income rural			Upper income urban Loc. IV
		Loc.[a] I	Loc. II	Loc. III	
		Pre-school children			
Urea N × 100/total N	N	26.0	29.0	18.0	22.0
	$\bar{x}$	60.0	70.8	72.9	82.9
	SD	13.3	10.0	9.6	6.8
Urea N/creatinine	$\bar{x}$	9.0	9.2	8.9	15.3
	SD	6.3	6.1	2.5	6.9
		School children			
Urea N × 100/total N	N	27.0	23.0	28.0	30.0
	$\bar{x}$	72.5	70.1	69.4	75.3
	SD	12.1	9.8	10.0	7.4
Urea N/creatinine	$\bar{x}$	7.0	6.3	6.0	12.0
	SD	3.5	1.8	1.3	8.0

[a] Locality; SD = Standard deviation.

Taken from: Arroyave, G., Biochemical characteristics of malnourished infants and children. *Proc. Western Hemisphere Nutrition Congress*; organized by the Council on Foods and Nutrition, American Medical Association, November 8—11, 1965, Chicago, Ill. AMA 1966, p. 30—36.

protein intake. Table 2 gives typical results obtained from the application of this method to a Central American population group. These results are consistent with findings in many other parts of the world (Platt, 1954; Luyken and Luyken-Koning, 1960; Couvée, Nugteren and Luyken, 1962). The rate of urine excretion has a large direct effect on the renal clearance of urea at low levels of protein intake, but not at high levels (Arroyave et al., 1966; 1967). This is shown in Table 3. Consequently, the usefulness of the urea/creatinine ratio as an indicator of protein intake is maximum at low levels of urinary flow. It is essential, therefore, that when this method is applied, efforts should be made to restrict the water intake of the children previous to the test. The values for this ratio corresponding to the level of protein intake per kilogram of body weight have been investigated in our laboratories under controlled conditions (Arroyave

Table 3. *Effect of water intake on the basal urinary excretion of creatinine and urea nitrogen of children with dissimilar dietary characteristics*

Measurement	Group of children	Water load (ml/kg body weight)					
		0		15		30	
		X̄	SD	X̄	SD	X̄	SD
Urine volume (ml/min)	SAP[a]	0.40	0.25	1.86	0.41	2.70	0.71
	Orphanage[b]	0.65	0.24	1.58	0.42	2.33	0.67
Creatinine (mg/24 h)[c]	SAP	236.00	66.70	245.00	53.90	295.00	101.80
	Orphanage	270.00	112.20	243.00	53.70	264.00	99.70
Urea N (gm/24 h)[c]	SAP	2.17	0.68	3.78	0.72	4.21	1.19
	Orphanage	3.73	1.46	3.18	0.56	3.54	1.10
Urea N (gm/gm creatinine)	SAP	9.26	1.56	16.07	4.60	14.96	4.39
	Orphanage	14.12	3.01	13.45	2.71	13.79	2.34

[a] San Antonio Palopo; 21 subjects.
[b] Orphanage; 20 subjects.
[c] Calculated from approximately 3 h collections.
SD = Standard deviation.
Taken from: Arroyave, G., Biochemical characteristics of malnourished infants and children. *Proc. Western Hemisphere Nutrition Congress*; organized by the Council on Foods and Nutrition, American Medical Association, November 8—11, 1965, Chicago, Ill. AMA 1966, p. 30—36.

et al., 1966). Fig. 5 gives some results of these studies which have recently been confirmed with investigations in an additional group of children. It may be possible, therefore, to predict the prevailing level of protein intake of a group of children from the estimation of their urea/creatinine ratios in urine. Further studies are, however, necessary to determine the effect of protein quality besides quantity, but it is obvious that a low urea/creatinine ratio indicates a low protein intake. The method presents less complications than a dietary survey.

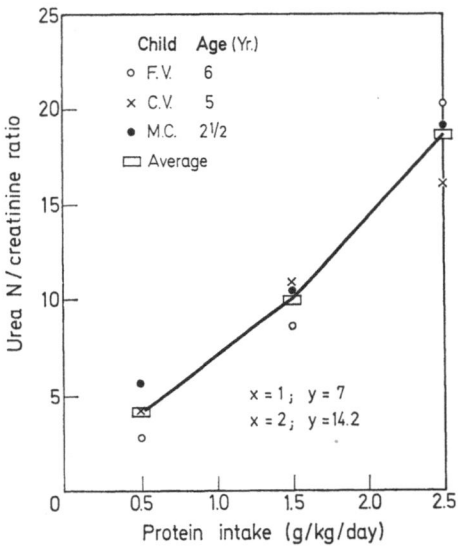

Fig. 5. Ratio of urea nitrogen to creatinine in fasting urine at controlled levels of protein intake

Discussion

We often hear statements, and even read them in scientific journals, comparing the relative merits of one or another biochemical measurement to evaluate nutritional status. This is particularly true about protein nutrition. The purpose of the present article is to emphasise that, in most instances, these comparisons are not valid because the different methods are providing information about different aspects of metabolism which are affected at different stages and different degrees of dietary protein deficiency. Faced with an initial shortage of dietary protein, the excretion of basal urea nitrogen will decrease and a low urea nitrogen/creatinine ratio may be the only characteristic change in that population of children. As a consequence of this, and also depending on the extent, homeostasis will become insufficient to maintain a normal free amino-acid pool and alterations in the plasma free amino-acid pattern ensue. A decrease in protein synthesis (true biological growth) will then become operative and the excretion of hydroxyproline may be reduced. A severe enough, long-standing situation of this nature will result in a failure of the body to maintain an adequate protein mass in relation to the height of the child, and the creatinine-height index will be proportionally less than that for adequately nourished children. The integration of the information obtained from applying these four approaches may permit a more rational judgment about the presence and magnitude of the protein deficit and its effects on the population of children under study at the time of the examination.

Emeritus Professor of Pediatrics, University of Zurich (Switzerland)

Has Malnutrition Only Bad Consequences?
What is the Definition of Health?

G. Fanconi

With 5 Figures

Contents

If in this paper I seem to be hypercritical or even pessimistic, may I insist that on the contrary I appreciate very much what the different international organisations have attained since their foundation, especially after the Second World War. As physician and as teacher I was and I still am an optimist. But I believe that it is always useful to consider problems from different standpoints.

The receiving capacity of developing countries is not very great; people are not interested in the American way of life, they are not able to cooperate especially to work, and frequently the young intellectuals do not know how to collaborate and are even arrogant.

In the populations of developing countries we have to do with a vicious circle of at least five links: poverty, ignorance (analphabetism), bad hygienic surroundings, malnutrition and diseases (Fig. 1). It is a rule that the help starts from the weakest link. It seems that poverty is the weakest one; the rich countries collect money which is sent to the poor, only a part of this money reaches the poor people. Either it gets lost in bureaucracy or it goes into the pocket of a few rich people in the poor countries.

The second weak link in the vicious circle are the diseases. "To cure a trachoma or a meningitis" wrote Dr. Berthet, the director of the Centre International de l'Enfance in Paris, "is not so difficult, but much more important is to improve the general hygienic conditions and the malnutrition, so that people would be less infected and more resistant".

Much more difficult is the struggle against ignorance and malnutrition. In many countries of Africa, primary and secondary schools have been introduced, but generally for the adolescents leaving the school there exists no possibility of occupation; for instance in Dahomey, only 20% of the young men leaving the school find an adequate occupation. Furthermore, they disseminate in their families what they have learnt at school and not always in a manner to improve the living conditions; for instance the son of a personality in Tunis told his sister that it is better not to nurse her child to conserve the line of her body! This advice, if generalised, increases the danger of weaning too early in a country where a good artificial infantile food is not easily available; the danger of infantile malnutrition is then high. We heard this morning from Dr. Jelliffe that too early weaning as a consequence of "better" education is today a real menace in the developing countries.

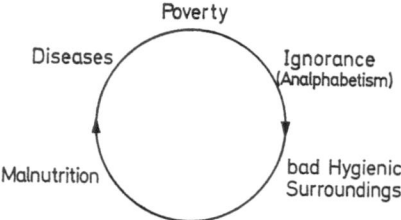

Fig. 1. Circulus vitiosus

Also the struggle against malnutrition is not at all easy. We know marasmus, the lack of calories, and kwashiorkor, the lack of protein, but what is the best, the healthy nutrition ? I do not believe that the European and American way of life with regard to nutrition is the best one for everybody. The definition of health by the WHO is not only the absence of disease, but a state of complete physical, mental and social well-being. I have not got the time to analyse this definition. May I only discuss a little *the definition of health from the nutritional standpoint*; malnutrition is multiconditional. Which are the criteria of good nutritional health ?

1. The *weight:* At the end of the Second World War the population in Switzerland was perhaps hungry and light (my weight was 68 kg, instead of 78 to 80 kg but I could work very well). Diabetes mellitus and dental caries became rare diseases. In many regards we were healthier than today. Furthermore Brock (1961) (Capetown) could demonstrate that malnourished people are better resistant to acute hunger than well-nourished. People suffering from vitamin-E deficiency have a decreased capacity to reduce methemoglobin to hemoglobin in their erythrocytes by the action of glucose-6-phosphate dehydrogenase and glutathione reductase, but if the deficiency is not complete they do not develop

a hemolytic anemia but are more resistant to malaria and other parasites because the parasites prefer to live in healthy red cells. In this regard people suffering from E-hypovitaminosis are healthier than the normal (Binder). Furthermore I know people with some digestive troubles; they have to be careful with eating, therefore they are protected against obesity and other troubles of overnutrition. In this way the digestive "disease" improves the general health.

2. The *body length:* Are people with secular acceleration of growth, probably caused by better nutrition especially with vitamins, healthier? They frequently present muscular weakness and troubles of the vertebral column (Scheuermann disease etc.).

3. *Resistance* to diseases. On the one hand, we know that a good nutritional state increases resistance to bacterial infections but on the

Table 1. *Proportions of cereals and starchy roots and animal products in different parts of the world*

	Cereals and starchy roots %	Animal products %
North America	25	40
Great Britain	30	30
Latin America	54	17
Africa	66	11
Near East	71	9
Far East	73	5

After N. Wright (1967).

other, it decreases resistance to viral infections, e.g. against the polioviruses. The virus prefers well-nourished cells as host.

4. *Happiness:* Well-nourished people are more aggressive and therefore less happy. On the other hand malnourished people have a lower level of working activity, they prefer the "dolce farniente" of the Napolitans. Highly civilised people, well-nourished, socially secure, grow up without any effort, without any privation, but effort and the struggle for life are necessary for happiness. The revolution of our students is a demonstration that, if the impulse of aggression is not satisfied and cannot be unloaded in another way, such as sport, then, in the dehumanised life of the city, it may produce very curious phenomena, e.g. student revolution.

5. *Malnutrition* may be the cause of many genetic diseases being eradicated before the bearer can have children and transmit them the pathologic gene. On the other hand, bearers of genes which become manifest only in well-nourished people, e.g. diabetes mellitus, may multiply more easily in malnourished populations.

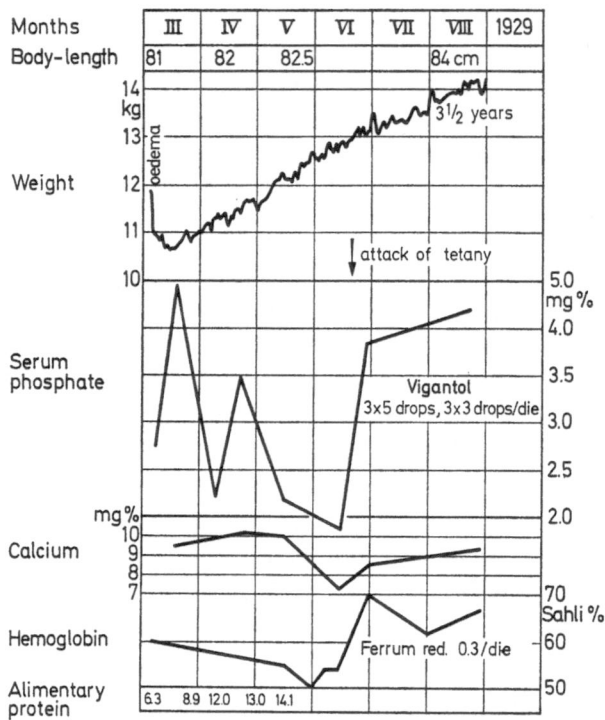

Fig. 2. Characteristic features of a classical case of coeliac disease in a child during recovery by feeding pure vegetable protein diet, devoid of gluten

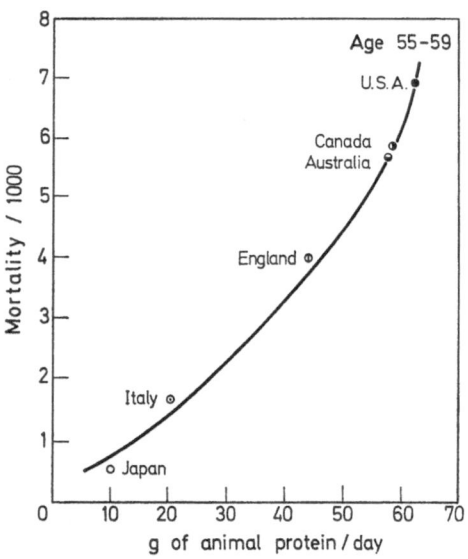

Fig. 3. Seeming dependence of the degenerative heart diseases of the older man upon the quantity of animal protein in the food (Holt, L. E., 1960)

May I make two further remarks concerning our problem:

1. In Japan (Table 1) the food is very poor in animal proteins and nevertheless kwashiorkor is not seen there (Katsunuma, Professor of Public Health, Tokyo, 1967). Furthermore, 40 years ago I believed that animal proteins could be the cause of the coeliac disease and I had very good results with a pure vegetable diet poor in proteins; some children recovered very well not because we omitted the animal proteins but because we gave a diet without gluten (Fig. 2).

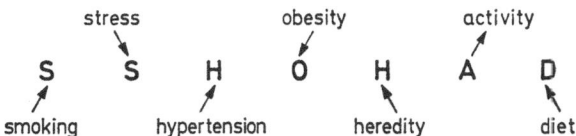

Fig. 4. The most important causes of cardiovascular diseases
(Eisenberg-Evang, 1967)

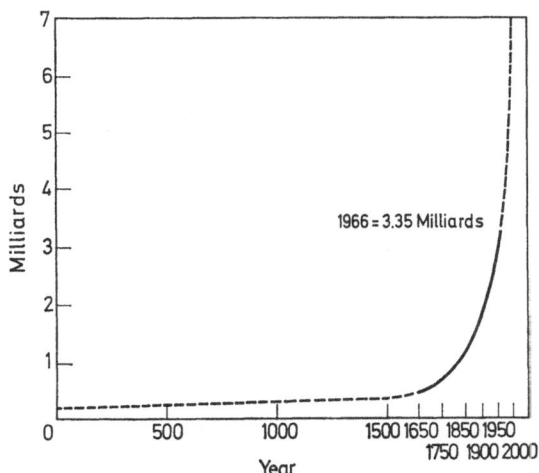

Fig. 5. The population explosion since the beginning of our century

It is statistically proven that the frequence of cardiovascular diseases at the age of 50 to 60 years increases with the intake of animal protein in the food (Fig. 3).

But such a statistical parallelism of two disparate phenomena is never proof of a direct causal relationship. With the increase of protein intake many other phenomena change and they may be more important causes of cardiovascular disease. In Fig. 4 are indicated the most important causes of the cardiovascular diseases.

2. The population of the world (Fig. 5) is increasing more rapidly than the production of food. We have to do with "une multiplication dans

la misère". The U.S.A. and Canada produce so much food and in such a cheap way that they can sell it at a dumping price. Therefore, e.g. in Nigeria (Prof. Lambo, 1967), it is cheaper to import American food than to produce it locally. Also in Switzerland, dumping prices and protection of the farmers, especially in the mountains, can never be combined.

I mention all these aspects not to say that the struggle against malnutrition is nonsense, but to consider how many difficulties arise in our task.

From the Caribbean Food and Nutrition Institute, Kingston (Jamaica)

Observations on Protein-Calorie Malnutrition in the Caribbean

D. B. Jelliffe

With 1 Figure

Contents

The classical picture of protein-calorie malnutrition (PCM), as seen for example in more traditional parts of Africa, has its peak incidence in the second and third years of life when kwashiorkor, in its classical form, is the main severe syndrome.

This situation is still widespread in some parts of the world, but in the English-speaking Caribbean a very different picture is seen. In this area, PCM has, as it were, "moved to the left". Possibly as a result of urbanisation and a failure of lactation, PCM occurs much younger than was the case in the classical descriptions in Africa.

It seems that the trend in the Caribbean may be of considerable importance, because it is probably the future pattern of other developing countries in which urbanisation is occurring. Incidentally, it may well also be very similar to the picture seen in the early days of the Industrial Revolution in Europe.

The Main Features of the Pattern of Protein-Calorie Malnutrition in the Caribbean are as follows

1. Severe syndromes are for a very large part infantile, i.e. occurring in the first year of life. For example, *kwashiorkor* occurred in Trinidad with its main incidence between the ages of 5 and 7 months (Jelliffe D. B., Symonds, Jelliffe E. F. P., 1960). The etiology was essentially that children were weaned from the breast very early and were fed on largely carbohydrate gruels, the most dangerous of which was arrowroot.

The clinical features of infantile kwashiorkor are rather different from those of the classical picture seen in the second and third years of life. In particular, marked hair changes are uncommon and skin lesions are not as frequent, while considerable enlargement of the liver is seen in the majority (Jelliffe D. B., Symonds, Jelliffe E. F. P., 1960; Waterlow, 1948).

The possible late effects of infantile kwashiorkor are extremely important because it may well be that severe malnutrition occurring in such young children leads to greater potential damage to the central nervous system, which is currently regarded as being much more vulnerable at this early age.

2. More common than kwashiorkor in the present-day Caribbean is *marasmus*, very frequently associated with diarrhoeal disease. So frequent is this association that it may often be better to term the whole condition "marasmus-diarrhoea".

This is the commonest cause of admission to hospitals and the most frequent cause of death in young children (McKenzie, Lovell, Standard, Miall, 1967). Hospitals statistics suggest that this condition is increasing in incidence, as opposed to kwashiorkor, which may be becoming less common.

3. An aspect of PCM, which has been under-emphasised in the world and certainly in the Caribbean as well, is the "bottom of the iceberg", the mild and moderate forms of PCM. Recent work in the Caribbean has shown that *mild-moderate protein-calorie malnutrition* is extremely common (Fox, Campbell, Morris, 1968); for example, in the island of St. Vincent, a recent field survey demonstrated that about one-third of all young children showed moderate protein-calorie malnutrition as judged by anthropometric measurements, particularly weight-for-age and arm circumference-for-age below the 3rd percentiles.

The Etiology of Protein-Calorie Malnutrition

In the Caribbean, infectious diseases act as conditioning factors, but with rather less importance than in Africa or in Central America. Malaria is non-existent, at least in the English-speaking islands; measles and whooping-cough have very much less nutritional impact than elsewhere. The two main infective conditions which are relevant are diarrhoeal disease, as has already been mentioned, and tuberculosis. Protein-calorie malnutrition in the Caribbean is therefore principally related to a *defective pattern of infant feeding:*

The main components of this defective pattern are:

1. a decline in lactation performance,
2. a rise in the practice of inadequate artificial feeding,
3. early and late transitional foods defective in both protein and calories.

Moulding Forces and Current Pattern of Infant Feeding in the Caribbean

There are seven moulding forces which influence the mother or restrict her choice with regard to the foods which she is likely to give to her young child:

1. Economic Constraints

People in the Caribbean have a low purchasing power, although not of the same level of poverty as in some other areas of the world. There is underemployment, seasonal employment and unemployment. All of this means that relatively little money will be available for the purchase of food, specially more expensive protein foods.

2. Defective Food Production

Local food production is much less needed in the Caribbean, especially protein foods such as those suitable for young children. This is partly related to the historical pattern in the area, in that the traditional agriculture was related to the production of cash crops on plantations for export to the metropolitan country rather than foods for local consumption.

3. Family Pattern

Illegitimacy is the norm for the majority of the less well-to-do segment of the community, and much more important than this, "families" tend to be matrifocal single parent units, i.e. the biological fathers often assume little responsibility for caring for or providing for their young children. The burden for this falls on the mother and on the grandmother. This has immense consequences as far as child-rearing and child-feeding are concerned.

It means, for example, that the mother will have to care for the child, will have to find the money to purchase food for the child, and also, while she is working, will have to find somebody who can look after the child while she is away. In particular, this poses great problems with regard to breast-feeding.

In addition, large families and what may be almost termed "a cult of fertility" is still an important cultural characteristic. The impact of this on intra-familial nutrition and also on general national nutrition is quite apparent. The rising tide of mouths to be fed is constantly out-pacing family incomes, national food production and the development of social services.

4. "Traditional" Practices

As all over the world, traditional practices exist, but in the English-speaking Caribbean it is very difficult to define what is meant by "tradi-tional" because, although the majority of the people are derived from West Africa, they came from different groups there, from different peo-ples, and they were then exposed to the experiences of slavery, during which time traditional tribal practices were, to a considerable extent, erased, mixed or altered. Nevertheless, what we may term "traditional practices" nowadays exist. As anywhere, some of these are beneficial, some are merely "neutral", and some are harmful.

In particular, one may mention the concept, in some of the islands in the Eastern Caribbean, of the milk-bag. In this, it is believed that a child on milk, whether it be breast milk or cow's milk, has in its stomach a bag of curdled milk which has to be removed before he can start on a wider diet. This is accomplished by means of restrictions of diet, laxatives and other medicines. The effect on the child is certainly harmful. Possibly this idea may have derived from people in the past having seen a calf after it had been slaughtered, in which case they may have been im-pressed by the fact that the stomach was full of curdled milk.

5. Out-Moded Imported Nutrition Education

Much of the nutrition education given in the English-speaking Caribbean is valuable, but at least some is unfortunately that of the United Kingdom of some 30 years ago, and is therefore both out of date and also locally irrelevant. For example, the use of cod-liver oil for young children in islands where rickets is non-existent, and the advice that babies should be given orange juice when in fact they are receiving ascorbic acid through their mother's milk.

6. Commercial Persuasion

Widespread and totally inappropriate advertising of excellent, but impossibly costly, imported canned milk, infant foods, and highly misnamed "tonic foods" is a major factor in influencing a mother in the English-speaking Caribbean towards purchasing of foods which she believes best for her young child. This poses major problems because, in fact, with the *per capita* income of the less well-to-do majority, it is completely impossible for mothers to purchase adequate quantities. In addition, the question of preparing milks with a very limited equipment, with a three-stone kitchen, a single pot, and no store, and with a dirty environment means that it is extremely likely that a contaminated, over-dilute feed will result and that the child will be started on the road to diarrhoeal disease and marasmus.

7. Imitation of Socio-Economic "Superiors"

In socially highly stratified societies such as found in the Caribbean, upward mobility may be difficult to achieve and frequently one way open to less well-to-do people is "eating their way up to the socio-econo-mic pyramid". This may often lead to an improvement of dietary pat-terns, but also the opposite may occur, and in the Caribbean and else-where, the taking-on of the practice of bottle-feeding with cow's milk rather than breast-feeding may be particularly deleterious. The lactation failure seen increasingly in urban regions and sometimes even in rural regions in developing parts of the world is related to two main features:

a) mothers may abandon breast-feeding (except at night), because they may have to go out to work;

b) equally often, there may be lactation failure because of interference with the psychosomatic let-down reflex. This let-down reflex is basically responsible for producing ejection of milk from the alveoli into the terminal lacteals, as a response to suckling stimulation.

However, breast-feeding is a "confidence trick", and the let-down reflex is enhanced by an untroubled confident mother. Conversely, if there is anxiety and doubt, then interference can occur with the let-down reflex and one gets the vicious circle leading to lactation failure seen only too frequently in Western countries and, more recently, also in less developed regions.

Conclusion

The approaches to improving the *community nutrition level*, especially its more vulnerable segments, particularly young children, has to be envisaged from a wide perspective. There is no single or simple approach to the problem. At the Caribbean Food and Nutrition Institute, we use a non-mathematical equation as a guide to our thinking (Fig. 1).

In this, we envisage the community nutrition level as being related to interactions between the economic and education levels, food availabili-ty, and aspects of health, all divided by the universal denominator of population size. Time does not permit us to look into the different aspects of this equation, but if one considers methods of improving

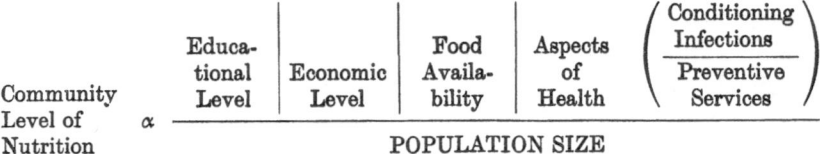

Fig. 1. An expression of the community level of nutrition

infant feeding practices alone, this, in the Caribbean, is largely related to a process of supplying effective *education* and *information*. Three main groups of persons are concerned or should be concerned in this process, and they are as follows:

1. Administrators, Policy Makers, Politicians

They should be made aware of the problem of the existence of protein-calorie malnutrition in the community. Frequently they are ignorant that these problems exist in their particular country or region.

Secondly, there is a need for the development of a rational food and nutrition policy with collaboration between people from different ministries and disciplines.

2. Would-be Nutrition Educators

Too frequently in the past, and indeed in the present, those responsible for nutrition education in the health field or in other fields have been ill-prepared to adapt their message to actual local problems and circumstances. Too frequently, it has been an attempt at the transportation of practices from one part of the world to another. Nutrition education has to be relevant to the ecology, to the culture pattern, to the kitchen and also always has to bear in mind the question of cost.

3. Food Industry

At the present moment, the food industry of the Western world is doing a considerable amount of harm in promoting the sale of its excellent, highly advertised, costly and locally inappropriate infant foods in less developed regions. An important question is as follows:

Is it justifiable to employ modern high-powered advertising techniques to sell infant foods in regions where there is no possibility of the vast majority of mothers being able to purchase adequate quantities or carry out the techniques satisfactorily? Particularly, is this so if there is still a pattern of successful lactation?

The real need is rather for the food industry to concern itself with what is really required, commercially produced, potentially profitable, low-cost, high-protein, high-calorie, high-nutrient, high-prestige foods, preferably prepared largely from local foods and introduced to suit local needs and ecological circumstances, and to avoid the interference with the local lactation pattern.

However, it has also been realised that the need for these types of food is by no means universal. In some rural areas, they may not be feasible because of low purchasing-power.

In some areas, the following three types of situations may have to be borne in mind by those concerned with the food processing:

a) artificial feeds for the occasional baby who cannot be breast-fed for various reasons, such as the mother going to work or having died or being seriously ill;

b) the treatment of severe protein-calorie malnutrition in hospitals. This may be the same food as (a);

c) transitional foods. Various forms of high-protein, high-calorie transitional foods may be indicated. They may be prepared for the young child alone or as a family food of which the young child may partake. They may be in the form of an additive or a complete food. Basically, there is a need for the realisation that both health personnel and others engaged in nutrition education, and also those in the food industry, have not been carrying out their function in the most relevant fashion for less developed regions in the past, and there is a need for them to re-appraise what they are doing and to look together at the actual needs.

Primarily, it is necessary that the three groups of people referred to, the administrators, the nutrition educators and the food industry, should re-examine their ideas. Without this process of re-thinking by these groups, it seems unlikely that truly relevant nutrition education can be carried out on the feeding of young children in less developed regions.

From the Departments of Biochemistry and Nutritional Sciences, University of
Wisconsin, Madison (Wisconsin, U.S.A.)

Amino-Acid Imbalances: What is their Significance in Relation to Protein Deficiency?

A. E. Harper

With 4 Figures

Contents

The term amino-acid imbalance is used to characterise an experimental condition in which the rate of growth of an experimental animal is
depressed by a supplement of an amino acid or acids other than the one(s)
that is most limiting in the diet, with the proviso that none of the amino
acids in the supplement is present in an amount that can be considered
toxic. Amino-acid imbalances are most readily created in diets that are
low in protein, and the growth depression due to an amino-acid imbalance
is readily and completely prevented by a supplement of the growth-
limiting amino acid.

Two types of amino-acid patterns characteristic of diets having
amino-acid imbalances are illustrated schematically in Fig. 1. A number
of examples of amino-acid imbalances created by adding small amounts
of the second limiting amino acid(s) to a diet that provides about half
the required amount of protein have been reported (Harper, 1958). Such
a procedure does not routinely result in depression in growth rate. A
reliable and reproducible method of creating an amino-acid imbalance
consists of adding a mixture of indispensable amino acids devoid of the
limiting amino acid to a diet that is low in protein. Both types of imbalance are corrected by a supplement of the limiting amino acid.

The second procedure has been used in our laboratory to provide a
reproducible model for studies of the nutritional and metabolic effects of
amino-acid imbalances in the white rat. We have used diets containing
6% of casein supplemented with methionine and an equal amount of an
amino-acid mixture devoid of threonine to create an imbalance in-

volving threonine or 6% of casein supplemented with methionine and threonine and an amino-acid mixture devoid of histidine to create an imbalance involving histidine. The effects of these imbalances are most readily demonstrated in young animals with a high growth potential.

It is important before going further to distinguish clearly between an amino-acid deficiency and an amino-acid imbalance. Effects of an amino-

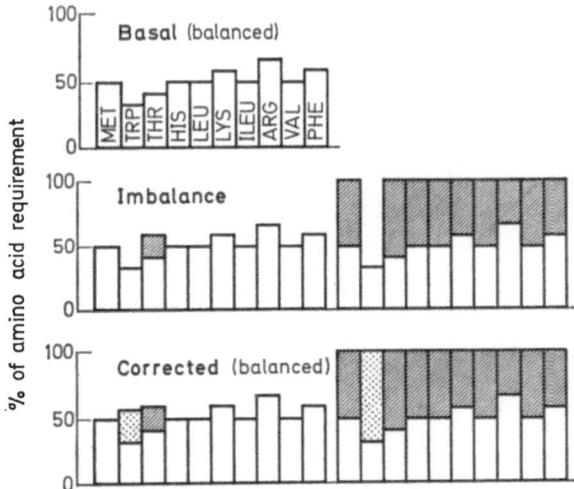

Fig. 1. Amino-acid patterns of diets used for the study of amino-acid imbalances

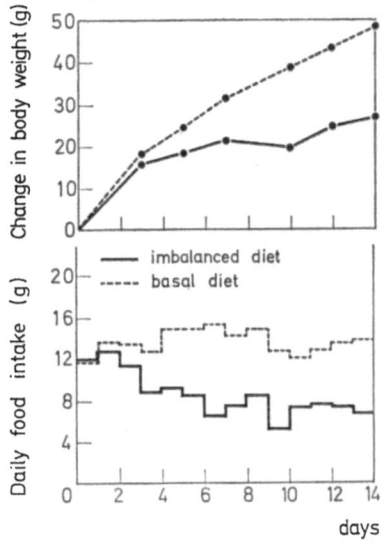

Fig. 2. Food intake and growth rate of rats fed basal or imbalanced diets

acid deficiency are attributable to a lack of or inadequate amount
of one or more indispensable amino acids. Effects of an amino-acid
imbalance are attributable to a surplus of amino acids other than the one
that is limiting for growth. The diets used in experimental studies of
amino-acid imbalances are usually deficient in at least one amino acid
but no more so than the control diet.

Growth rates and food intakes of rats fed the 6% casein-methionine
control diet or the diet with an imbalance involving histidine (control
diet to which a mixture of indispensable amino acids devoid of histidine

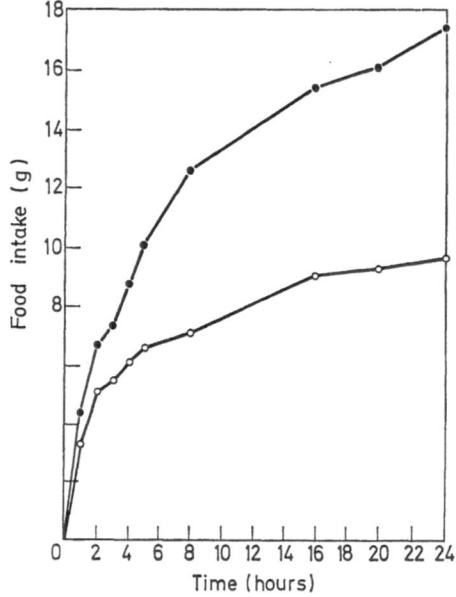

Fig. 3. Food-intake pattern of rats fed basal or imbalanced diets

has been added) are shown graphically in Fig. 2. The amino-acid imbalance depressed growth rate and food intake. The depressions occur
on the first day if the animals are not protein-depleted but are frequently
delayed for 1 to 3 days in protein-depleted animals.

Since a supplement of histidine prevented the growth depression and
stimulated food intake, we assumed initially that the extra histidine was
required because the imbalance reduced the efficiency of utilisation of
the limiting amino acid, possibly by stimulating its oxidation. However,
nitrogen balance studies indicated that efficiency of nitrogen utilisation
was depressed as much, if not more, in rats fed the control diet by
restriction of their food intake to an amount equivalent to that consumed
by the rats fed the diet with the imbalance.

Depressed food intake owing to an amino-acid imbalance of this type can be detected within a few hours as shown in Fig. 3. In this experiment rats were kept without food for 18 hours and were then offered the control diet or the diet with the imbalance. Food-intake measurements were made at intervals during the subsequent 24 hours. Observations of this

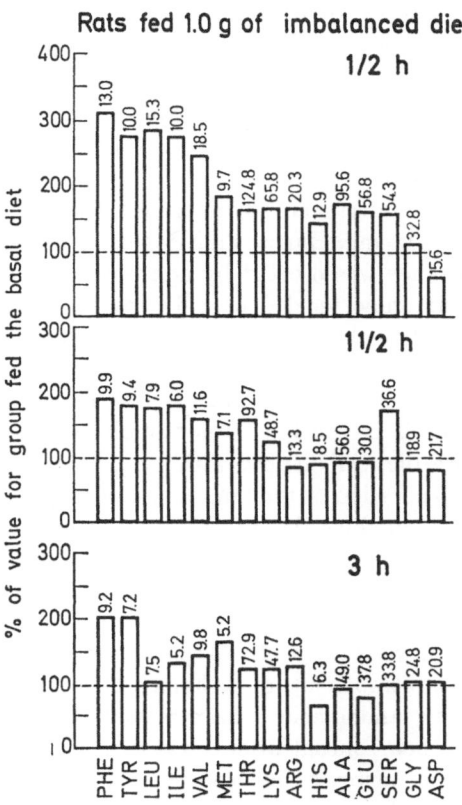

Fig. 4. Plasma amino-acid patterns of rats at intervals after ingestion of a single meal of a diet with an imbalance involving histidine. The dotted line represents the control value

type suggested that, whatever the metabolic effects of an amino-acid imbalance were, they must occur very shortly after ingestion of the diet, and that the growth depression was the result of food-intake depression (Harper, Leung, Yoshida and Rogers, 1964; Harper, Rogers, 1965).

Examination of plasma amino-acid concentrations of rats fed a diet with an amino-acid imbalance has revealed that the depression in food intake is associated with elevated concentrations of the amino acids added to the diet to create the imbalance and a depressed concentration of the limiting amino acid. The type of plasma amino-acid changes observed at intervals after the rat has ingested a single meal of a diet with an imbalance involving histidine are illustrated in Fig. 4. This type of plasma amino-acid pattern resembles that observed in animals fed a diet deficient in a single amino acid, a condition that also results in rapid and severe depression of food intake. Since pathologic changes occur in several tissues and survival time is reduced when rats are force-fed a large amount of a diet devoid of one amino acid (or of a low-protein diet) the possibility has been considered that the depressed food intake of rats fed either amino acid-deficient or imbalanced diets is a protective response, probably initiated by some signal associated with the altered plasma amino-acid pattern.

Isotopic studies undertaken to determine the fate of ^{14}C-labelled threonine (Yoshida, Leung, Harper and Rogers, 1966) in rats fed a diet in which an imbalance was created by addition of a mixture of amino acids devoid of threonine yielded no evidence that the imbalance stimulated oxidation of the limiting amino acid (Table 1). There was, however, increased retention of radioactivity in the carcass, much of it in the liver. Further studies (Benevenga, Harper and Rogers, 1968) indicated that protein synthesis in the liver was stimulated within a few hours after ingestion of a diet imbalanced with respect to histidine (Table 2). The enhanced incorporation of the limiting amino acid into liver proteins would be expected to reduce the amount circulating in body fluids and could account for the low concentration observed in the blood.

The observations on the effects of amino-acid imbalances and other observations on effects of a high protein intake (Anderson, Benevenga, and Harper, 1968) have led us to the hypothesis that alterations in food intake represent a component of a homeostatic mechanism for the regulation of plasma amino-acid concentrations. If amino acids are in excess in the diet, their concentrations in tissues can be reduced only by increased catabolism, increased incorporation into protein or decreased intake. In the animal fed a low-protein diet, growth rate is low, protein synthesis is low, and activities of enzymes of amino-acid catabolism are low (Harper, 1965, 1968); hence, any surplus of amino acids, such as the supplements added to create imbalances, cannot be readily or rapidly cleared from the body. Ordinarily a surplus of amino acids would accumulate in blood only after the needs for protein synthesis had been satisfied. It therefore seems possible that elevated plasma amino-acid concentrations or something related to them could serve as a signal leading to curtailment of eating. A low concentration of one amino acid may be a further signal. If such signals were elicited by a diet having an amino-acid imbalance and low in utilisable protein, an immediate depression of food intake would result and, hence, retardation of growth.

Thus, the primary effect of an amino-acid imbalance would be to depress food intake which, if prolonged sufficiently, would contribute to further protein depletion and caloric restriction of an animal already receiving an inadequate amount of balanced protein. In many of the animal experiments, adaptation to imbalanced diets is observed with time, as indicated by a gradual improvement in food intake and growth rate if the imbalance is not too severe.

Table 1. *Distribution of radioactivity after feeding rats basal or imbalanced diets containing U-C¹⁴-threonine*

	Control (6% casein)	Imbalance (6% casein + 10% A.A. Mix. — Thr.)
	% of ingested C^{14}	
Expired CO_2	18.4	14.3
Urine	2.1	2.2
Feces	1.6	1.2
Carcass	70.1	74.6
Liver	5.9	7.2
Total	98.1	99.5

Each rat was fed 7 g of diet containing 8.5 μc of C^{14}-threonine at the beginning of the experiment and the same amount of the same diet without C^{14}-threonine 24 h later. Experiment lasted 48 h.

Table 2. *Effect of amino-acid imbalance on incorporation of radioactivity from limiting amino acid into liver protein*

	Radioactivity in liver protein dpm/mg N
A) Control (6% casein)	20 485
B) Imbalance (A + 6% A.A. mix-his)	32 157
C) Corrected (B + 0.1% L-his.HCl)	38 165[a]

[a] Corrected for dilution by extra dietary histidine. Values are significantly different from each other.

Unfortunately it is difficult to study the problem of amino-acid imbalance in natural diets and to distinguish the effects from those of a simple amino-acid deficiency, owing to the difficulty of removing specific amino acids from the diet. However, work with the experimental models indicates clearly that removal of the amino acids that cause imbalances results in both food intake and growth stimulation. Thus the possibility must be considered that, where low-protein diets are consumed by the child, a severely unbalanced dietary amino-acid pattern may lead to curtailment of food intake, and hence to curtailment of both protein and caloric intake.

From the Nutrition Research Laboratories, Hyderabad, India

Observations on Some Epidemiological Factors and Biochemical Features of Protein-Calorie Malnutrition

C. Gopalan

Contents

Epidemiology of Kwashiorkor and Marasmus

As we are all aware, two major forms of protein-calorie malnutrition, namely kwashiorkor and marasmus, have been recognised. The crucial question of practical significance is: are these two forms of protein-calorie malnutrition attributable to two entirely different types of dietary situations and do they, therefore, call for two entirely different methods of approach with regard to prevention and control? or do they represent two facets of one and the same disease with the same dietary background?

On the basis of experimental studies, it has been postulated that a deficiency of protein with adequate or more than adequate calories would lead to kwashiorkor, while a deficiency of calories would lead to marasmus. If this is true, then in most developing countries we are faced with two vast and entirely different public health problems calling for two different approaches for prevention. In such a case, the dietary supplements which would be needed to prevent and control marasmus would be different from those which would be necessary for the prevention of kwashiorkor.

On the other hand, if the two syndromes are different clinical manifestations of the same type of dietary inadequacy, then we are dealing with one problem capable of solution by the same approach. In that event, the current discussions on the relative importance of kwashiorkor and marasmus, or the detailed studies of the subtle biochemical differences between these two syndromes, would not have much practical significance. It may be expected that epidemiological studies would provide direct evidence which would throw light on this question.

a) Cross-Sectional Survey of Pre-School Children

An intensive survey, covering over 3000 children below 5 years of age, was recently conducted by the Nutrition Research Laboratories, Hyderabad, among the poor rural communities of South India. In a representative sub-sample of this group, a careful diet survey was carried out by trained investigators with a view to assessing the actual diets consumed by these children. These children belonged to the poorest sections of the community among whom kwashiorkor, marasmus and marasmic kwashiorkor have been found to be frequent. The study showed that the

Table 1. *Protein and calorie intake of poor South Indian children*

Age group	Number surveyed	Mean weight kg	Protein intake g/24 h	g/kg	Calorie intake kgcal/24 h	kgcal/kg
6—12 months	126	6.7	12.5	1.9	550	82
1—2 years	418	7.8	14.0	1.8	610	79
2—3 years	328	9.1	19.8	2.2	860	96
3—4 years	394	10.7	21.2	2.0	910	86
4—5 years	578	12.4	20.0	1.6	900	73

dietary pattern of children who developed kwashiorkor was in no way different from the dietary pattern of children who developed marasmus (Table 1). They all subsisted on the same type of cereal-based diets.

Clearly, there was no evidence whatsoever that the children who developed kwashiorkor had been force-fed or that their mothers tended to "push" starchy foods. The protein intake varied from 1.9 g/kg bodyweight in infancy to 1.6 g/kg in children 4 to 5 years of age — levels which would not appear inadequate according to the latest recommendations of the FAO/WHO Expert Group. The calorie intake, on the other hand, was around 80 Calories/kg in most age groups — a level which could be considered to be clearly inadequate. The main bottleneck in the dietary situation appeared therefore to be calorie inadequacy.

It would then appear that, at least as far as this community is concerned, the elegant hypothesis that marasmus is due to predominant caloric deficiency, while kwashiorkor is a result of predominant protein deficiency, will not hold.

b) Longitudinal Study of Pre-School Children

A longitudinal study covering 300 children right from birth for periods extending up to 3 years confirmed the above results. It was noted in this study that some of the children who developed marasmus at one point of time, on being followed up, developed a picture of kwashiorkor. Other children who had developed kwashiorkor later presented a picture of marasmus. Thus, marasmus and kwashiorkor were not only coexisting in the same community, but they were frequently seen in the same individual in different points of time. I have discussed my hypothesis with regard to the pathogenesis of kwashiorkor and marasmus in the light of these observations elsewhere and I feel that, in view of the accumulating evidence from different parts of the world, our earlier concepts regarding the pathogenesis of kwashiorkor and marasmus need reappraisal. Such reappraisal will not be a mere academic exercise but will have far-reaching practical implications.

c) Family Size and Incidence of Protein-Calorie Malnutrition

Recent epidemiological studies carried out at our Laboratories in Hyderabad indicate a significant relationship between family size and nutritional status of pre-school children. In the general population, children belonging to birth orders of three and below constitute 54% of the population. An analysis of the records of *all* children admitted to the out-patient department of our pediatric hospital showed that children belonging to birth orders below three accounted for 66% of the hospital admissions, as against 34% accounted for by children of the higher birth orders. The larger number of children of the earlier birth orders among hospital out-patients is partly a reflection of the relatively larger number of children of earlier birth orders in the community and partly an indication of the fact that mothers tend to seek hospital care more readily for their first few children.

An analysis of 872 cases of kwashiorkor investigated at the same hospital, however, showed that out of these children only 39% belonged to birth orders three and below, while 61% belonged to the higher birth orders. It is significant that, in spite of the general preponderance of the children of earlier birth orders among general admissions to the hospital, the great majority of cases of protein-calorie malnutrition consisted of children in the latter birth orders.

In another field study covering over 1400 pre-school children, it was found that, while 32% of children belonging to birth orders four and above exhibited various signs of malnutrition, only 17% of children of earlier birth orders showed such evidence. Even allowing for normal distribution of children in the two groups of the community, these data would show that 62% of all nutritional deficiency states in pre-school children are encountered in children of birth orders four and above.

This would indicate that, even under the present economic and living conditions, mere limitation of family size to three children can bring down the incidence of malnutrition in pre-school children in India by about 60%. This is perhaps an underestimate as it does not take into account the possible impact of country-wide family planning on the general economic status and food resources position in the country. I will now turn to a brief discussion of our recent studies in two important bio-chemical aspects of protein-calorie malnutrition.

The Skin in Kwashiorkor
1. Collagen Nitrogen and Non-Collagen Nitrogen

Ever since the classical description of kwashiorkor by Cicely Williams (1935), it has been recognised that skin changes constitute an important and striking clinical feature of this syndrome. It is, however, surprising that the voluminous literature on the subject of protein-calorie malnutrition contains few detailed studies of the biochemical changes in the skin in this disease. Such studies may be expected to throw light on the precise significance of these skin changes.

The skin constitutes an appreciable proportion of the total body-weight — about 8% in the adult and probably more in the child. The crude protein content of the skin is nearly 22% and accounts for about $1/8$ of the total body protein. It is obvious, therefore, that in any study of disorders of protein metabolism, the skin must receive attention. Of the proteins of the skin, collagen, which constitutes 70% of the total nitrogen of the skin and is mainly responsible for the mechanical strength and stability of the skin, is especially important.

In the light of these considerations, the results of a recent study carried out in the Nutrition Research Laboratories, Hyderabad (Vasantha et al., to be published) on certain changes in the amino-acid composition of the skin in cases of kwashiorkor may be interesting. Nineteen children suffering from kwashiorkor with ages ranging from 1 to 5 years, with and without classical skin changes, were included in this study. Specimens of the skin were obtained by biopsies at the time of admission and again after successful nutritional rehabilitation. A sample of skin measuring 2.5 × 5 mm was obtained from the anterior aspect of the thigh on admission and repeat biopsies after nutritional rehabilitation were carried out on an identical area in the opposite extremity. Biopsy specimens of skin were also obtained from 10 normal children between the ages of 1 and 5 years for purposes of comparison. A portion of the skin was taken for estimation of total nitrogen by the micro-Kjeldahl method. The rest was processed for dermal amino-acid analysis as described by Neldner et al. (1966). The dermis was separated from the epidermis by frozen section technique, dehydrated for 24 h at 105 °C and then continuously defatted for 8 h with ether in a Soxhlet apparatus. The defatted dehydrated bit of the dermis generally weighing between 2 and 4 mg was hydrolysed with 6 N HCl under vacuum at 110 °C for 24 h. After hydrolysis was complete, the acid was evaporated in a vacuum dessicator. The residue was dissolved in pH 2.2 buffer and analysed for its amino-acid composition, using the Spinco-Beckman automatic amino-acid analyser. An aliquot of the hydrolysate was taken for estimation of dermal nitrogen by the micro-Kjeldahl procedure.

Hydroxyproline × 7.09 = Collagen protein.

$$\frac{\text{Collagen protein} \times 18.6}{100} = \text{Collagen nitrogen.}$$

Total nitrogen — Dermal nitrogen = Epidermal nitrogen.

Dermal nitrogen — Collagen nitrogen = Non-collagenous nitrogen.

At the time of admission, the total nitrogen content of the skin of children with kwashiorkor was significantly lower than that of normal children of comparable age (Table 2).

This reduction was of a much greater magnitude in cases with skin lesions than in those without skin lesions. The major portion of this reduction was accounted for by reduction in the dermal nitrogen content. The skin showed a reduction in both collagen nitrogen and non-collagen nitrogen, the reduction in the former being much greater, with the result that the ratio of collagen nitrogen/non-collagen nitrogen was 1.5 instead of the normal value of 2. After treatment, the total nitrogen content of the skin as well as the ratio of collagen nitrogen/non-collagen nitrogen returned to normal values.

2. Amino-Acid Composition

In Table 3, I have indicated the results of our study of the amino-acid composition. In all cases of kwashiorkor, the hydroxyproline content of the skin was significantly lower than in normals when expressed either in terms of dry weight of the skin or as a percentage of the dermal nitrogen. The reduction was more marked in children with skin lesions than in those without. Following nutritional rehabilitation and disappearance of the skin lesions, there was a significant increase in the level of this amino acid.

The reduction in hydroxyproline content would clearly indicate a loss in collagen and, since this reduction was disproportionately greater than the reduction in total dermal nitrogen, it may be concluded that, among the various fractions of dermal proteins, collagen was preferentially lost.

Table 2. *Nitrogen content of skin in normal and kwashiorkor children*

Group	Total nitrogen[a]	Dermal nitrogen	Epidermal nitrogen	Collagen nitrogen	Non-collagen nitrogen	Collagen: Non-collagen nitrogen
Normal children (8)	18.7 ± 1.1	16.0 ± 1.4	2.7 ± 0.5	10.6 ± 0.7	5.2 ± 1.5	2:1
Kwashiorkor without dermatosis (11)	14.3 ± 1.2	13.0 ± 1.3	2.5 ± 0.8	7.7 ± 2.1	5.1 ± 0.8	1.5:1
Kwashiorkor with dermatosis (6)	10.8 ± 1.3	9.0 ± 3.3	1.7 ± 0.5	5.8 ± 1.7	3.1 ± 1.6	1.8:1
Kwashiorkor after treatment	17.7 ± 1.2	15.8 ± 1.0	1.7 ± 0.9	11.1 ± 1.3	4.4 ± 1.1	2.5:1

[a] Nitrogen values expressed as grams/100 g of dry defatted tissue.

Apart from the marked fall in hydroxyproline content, another striking feature was a reduction in the tyrosine levels. There is now evidence that tyrosine is related to the maturation and structural integrity of the collagen fibres. Tyrosine residues are known to be involved in the proper alignment of tropocollagen, maturation and fibril aggregation (Bensusan, 1960; Hodge, 1960; and Bowes et al., 1958). The fact that the reduction in tyrosine was particularly marked in children with "crazy pavement" dermatosis would appear to be specially significant.

These observations would confirm the speculation that the skin lesions in kwashiorkor are attributable to quantitative and qualitative changes in skin collagen. Apparently, for the development of crazy

Table 3. *Dermal amino acid levels expressed as percentage of dermal nitrogen*

Amino acid	Normal children (10)	Kwashiorkor without dermatosis (12)	Kwashiorkor with dermatosis (7)
Hydroxyproline	5.6	4.1[a]	4.0[a]
Serine	2.6	2.7	2.9[a]
Proline	9.6	8.7	7.9[a]
Glycine	26.9	25.1	19.9[a]
Tyrosine	0.75	0.76	0.62[a]
Arginine	6.3	11.8[a]	11.5[a]
Ammonia	5.5	10.0[a]	14.6[a]

[a] Difference is statistically significant.

Levels of the other amino acids — aspartic acid, threonine, glutamic acid, alanine, valine, isoleucine, leucine, lysine, phenylalanine were found to be not different from that of normal children.

pavement dermatosis, apart from reduction in the collagen content, a reduction in tyrosine content probably resulting in structural immaturity of the collagen is also essential.

Apart from hydroxyproline, proline and glycine also showed a significant reduction. This change is again indicative of reduction of collagen, since the contents of glycine, proline and hydroxyproline in collagen are known to be high. On the other hand, levels of arginine and ammonia were raised in cases of kwashiorkor; the precise significance of this change is as yet not clear.

The Thyroid Function in Protein-Calorie Malnutrition

It is now becoming increasingly clear that the stress of protein-calorie deficiency induces important hormonal changes which may be concerned in adaptation mechanisms and in modifying clinical response to the deficiency. Intensive investigation of the endocrinological aspects

of protein-calorie malnutrition is essential for a proper understanding of
the pathogenesis of kwashiorkor and marasmus. We had earlier reported
our observations (Jaya Rao et al., 1968) on plasma cortisol levels in
kwashiorkor and marasmus and the response thereof to synacthen
(synthesis β 1 to 24 Corticotropin).

I would like to present here some of our more recent observations on
thyroid function.

Kamala Jaya Rao Raghuramulu and Srikantia in the Nutrition
Research Laboratories have recently investigated thyroidal uptake of
iodine in 20 cases of kwashiorkor and in a series of normal children,
using ^{131}I (5 µc). In these studies serum protein-bound iodine (PBI) was
also estimated. These studies were repeated after nutritional rehabilitation in cases of kwashiorkor (Table 4).

Table 4. *Thyroidal uptake of ^{131}I and PBI in kwashiorkor*

	^{131}I uptake % of dose given		PBI µg/100 ml
	2 h	24 h	
Normal children	12.6	17.9	6.2
	(7)	(5)	(9)
Kwashiorkor:			
On admission	13.3	18.2	3.7[a]
	(12)	(12)	(22)
After therapy	17.0	20.1	6.0
	(5)	(5)	(18)

[a] 12 children had values below 3.5 µg/100 ml (mean 2.2).
10 children had values in the normal range (mean 5.4).

The percentage of tracer iodine uptake at the end of 24 h in kwashiorkor ranged from 11.3 to 29.6 with a mean of 18.2; nearly the same
values were obtained after treatment. The range in normal children was
from 9.8 to 28.7 with a mean of 17.9.

Thus, with regard to iodine uptake, there was no significant difference as between cases of kwashiorkor on the one hand and normal
children on the other; nor as between cases of kwashiorkor before treatment and the same cases after treatment.

Serum PBI values in kwashiorkor cases appeared to be considerably
reduced in some cases while they were nearly normal in others. In
children who had very low PBI values on admission (ranging from 1.3 to
3 µg per 100 ml), it was found that these values returned to normal levels
(5.1 to 6 µg per 100 ml) after treatment. In cases of kwashiorkor who had
initially a normal level of PBI, however, there was no significant change
in the PBI values.

These results would indicate that, while the uptake of iodine by the thyroid is normal in cases of kwashiorkor, the serum PBI levels may be reduced significantly in some cases. A reduction in serum PBI levels in the face of normal uptake would signify a block at some stage between uptake and thyroxin synthesis.

The first step in the hormone synthesis is the organic binding of iodine within the gland. A possible defect in this step was investigated through the perchlorate discharge test in 3 subjects who had low serum PBI levels and 2 with normal serum PBI levels. A tracer dose of iodine (5 μc) was administered orally and thyroidal uptake measured at the end of 60 min.

Immediately thereafter, 100 mg of potassium perchlorate were given orally and the thyroidal counts repeated 20 min and 60 min after the perchlorate was given. A discharge of more than 10% (which is considered to be positive) was obtained in all subjects who showed initially low PBI values but not in those with normal levels of PBI to start with. After treatment, the PBI levels were normal in all cases and thoses cases who had positive perchlorate discharge test initially returned negative results.

It would thus appear that a defective organification of iodine leads to low serum PBI levels in some cases of kwashiorkor.

From the Department of Research and Development of the Nestlé Group, Vevey
(Switzerland)

Single Cell Proteins:
Basic Aspects and Future Trends

L. Rey and J. Mauron

With 4 Figures

Contents

I am in a rather difficult position to discuss my subject in front of you,
who are specialists in the field of nutrition and clinical studies, since I am
basically a professor in biological physical chemistry; but nevertheless, I
wish to report on the studies done in the department of Research and
Development of the Nestlé Group by my colleagues and myself.

The production of microbial protein material is not new. Yeast has
been produced on various sugars and sugar wastes, on sulfite liquors etc.,
for over 50 years. However, the recent consciousness of the overwhelming
protein deficiency in the world has emphasised the need for thoroughly
investigating the possible applications of microbial protein as one solution
to this problem.

Since the protein content of most microorganisms is very high, the
latter can be used as such and protein has not to be extracted. The product consists therefore of whole cell bodies or fragments of cells and
has been named "Single Cell Protein" (SCP). It is of course possible to
extract the protein and to prepare it in a purified form (protein isolate)
but, in general, Single Cell Protein refers to the whole cell body and not
to the specific protein fraction.

The main points of my topic are best summarised with a series of
figures.

Table 1 shows the mass-doubling time for several organisms on a statistical basis. These are very well-known data. The mass-doubling time goes from 5 years for cattle down to 1 to 6 hours for bacteria, and it is obviously of very great interest to have a large mass of material in a short time if you are aiming at mass production.

Table 2 will give you an idea of the composition of the different proteins produced by biosynthesis in microorganisms. As you can see, bacteria are generally richer in proteins than yeast, moulds and algae on a dry weight basis. We shall restrict our comments for the beginning to bacteria and yeast.

Table 1. *Mass doubling times for several organisms, on a statistical basis*

Organism	Mass doubling time
Bacteria, yeasts and unicellular algae	1—6 hours
Higher plants (tubers, pulses, cereals)	1—4 weeks
Chickens	1 month
Pigs	4 months
Cattle	5 years

Table 2. *Composition of the different proteins produced by biosynthesis*

	Proteins % dry weight	Amino acids, % of proteins		
		Lysine	Methionine	Tryptophan
Bacteria	40—80	6.5	1.8	0.5
Yeasts	30—60	7.2	1.2	0.8
Moulds	10—40	6.8	1.3	0.8
Algae	40—60	5.2	2.0	—

Generally speaking, they are fairly rich in lysine, somewhat deficient in methionine and sometimes also in tryptophan.

Table 3 gives you a list of some microorganisms liable to use hydrocarbons. As you know, normal paraffins are available in very large amounts up to hundreds of million tons every year, and they can be transformed either by bacteria or by yeast into cell bodies. In fact, we have been concentrating our efforts in the last 3 years — jointly with our friends of Esso Standard Oil — to develop some process making use of hydrocarbon as a feed-stock to grow microorganisms and to produce Single Cell Protein.

This is just to give you an idea of the variety of the organisms liable to grow on hydrocarbon feed-stock.

Table 4 shows a lay-out of a plant for production of bacterial biomass from hydrocarbon substrate. From the beginning, we should understand that hydrocarbon is bringing only carbon and hydrogen as the composi-

Table 3. *Types of organisms utilising hydrocarbons*

Bacteria	Mycelial bacteria	Fungi (including yeasts)
Achromobacter	Actinomyces	Acremonium
Alkaligenes	Micromonospora	Aspergillus
Bacillus	Mycobacterium	Candida
Bacterium	Mycococcus	Citromyces
Brevibacterium	Nocardia	Debaryomyces
Corynebacterium	Streptomyces	Endomyces
Flavobacterium		Hansenula
Micrococcus		Monilia
Pseudomonas		Penicillium
Vibrio		Scopulariopsis
		Torula
		Torulopsis

Table 4. *Protein concentrate from heavy gas oil and n-paraffins*

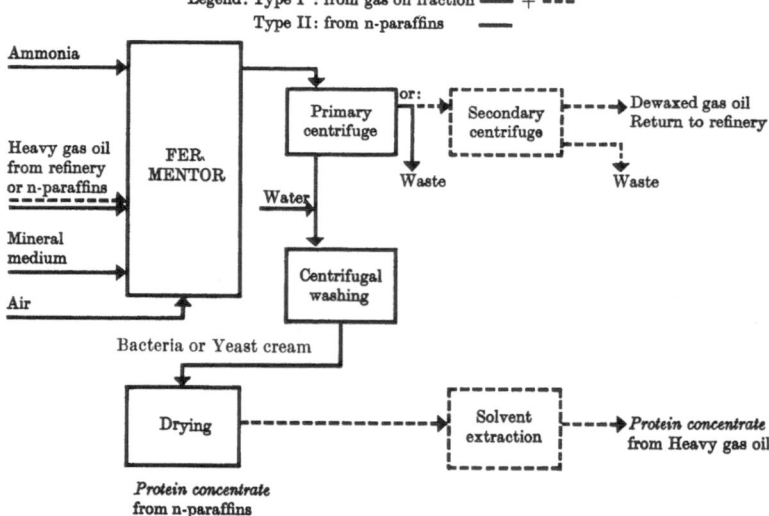

tion indicates. You have to feed oxygen for the fatty acid and carbohydrate synthesis and nitrogen in order to build the protein.

Very briefly said, two fermentors are used. One is for inoculation and the other one is for mass production. The latter fermentor is, of course, fed continuously with the substrates.

The main substrate is a hydrocarbon cut; it is generally normal paraffin of pharmaceutical grade in the C 12 to C 18 range. The aqueous medium, including mineral salts, ammonium salt and phosphoric acid, procures the vital elements for the growth of the microorganism. Since we are dealing with a system of four unmiscible phases (air, microorganism, water, hydrocarbon), vigourous stirring and the addition of an emulsifier are essential.

The product which is made in the fermentor consists essentially of the slurry of the cell bodies. It is extracted from the fermentor on a continuous basis, the cell bodies being centrifuged while the overflow is recirculated in the fermentor since we have a continuous process. We obtain concentrated live cells, which have to be sterilised by steam-heating prior to drying. The latter is performed in a conventional spray-

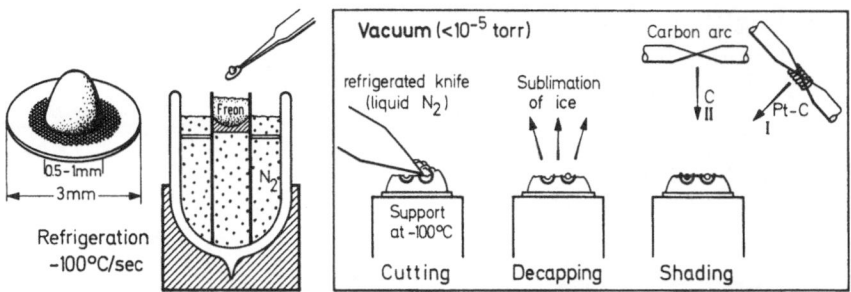

Fig. 1. Structure of different cells produced in fermentor

dryer. The grey powder thus obtained is of rather bland taste and has good keeping properties.

The lay-out for a plant producing, say, some 50 million to 100 million lbs of dry protein a year is now ready.

In fact, in our pilot-plant, which is now operating in the U.S.A., we have fermentors up to 2000 litres capacity working on a continuous basis. This means that we can already produce some large amounts of material for biological and clinical testing.

Fig. 1 illustrates some of the studies that we have been doing in order to investigate the structure of the different cells produced in the fermentor.

When using microorganisms as a food, we encounter some difficulties due to the tough cell membrane. A thorough investigation of the structure and composition of the cell membrane is therefore a must. So we have started not only to isolate the cell membrane by differential centrifugation, but also to study its intimate structure.

For this purpose, we have been using the freeze-etching technique developed at the Polytechnical Institute of Zurich. This technique consists in freezing the material first at —100°, then etching the surfaces by a superficial freeze-drying for a period of minutes, and then making a replica of this by carbon and platinum. This replica is examined under the electron microscope.

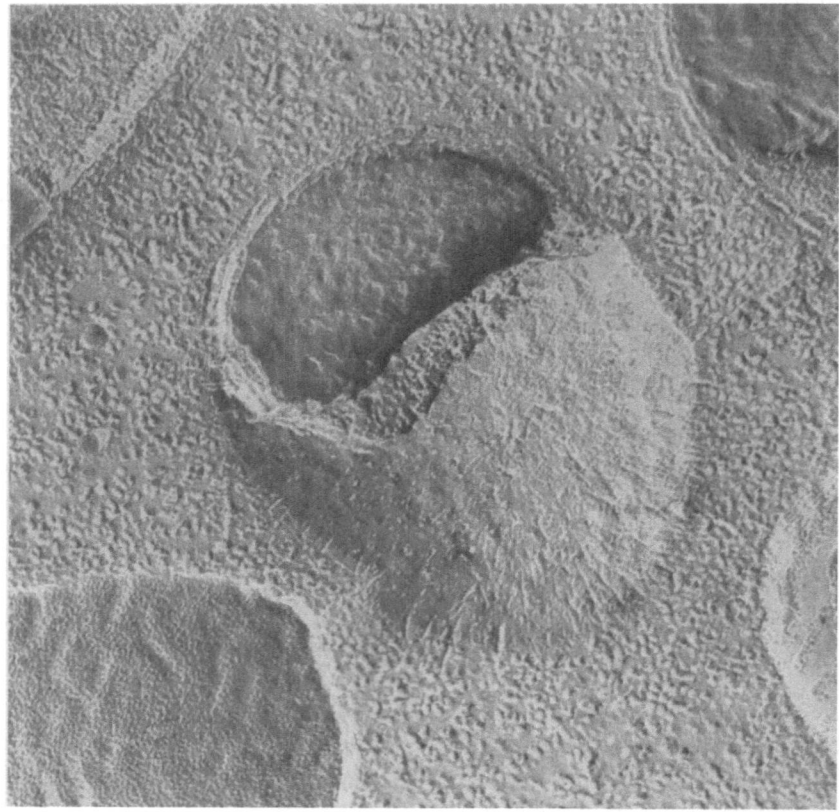

Fig. 2. Freeze-etching technique: Bacterial cell

Fig. 2 represents a bacterial cell; you can see the different layers of the membrane, as well as the inside of the cell structure.

Fig. 3 is obtained from a dividing yeast cell. One can very well see the nucleus and the communication between the peri-nuclear space and the nucleus itself, with some vacuoles inside. One can also see the different layers inside the membrane itself.

Leaving the structural studies, we may now give you a brief account of the biochemical aspects of the problem.

Table 5 shows the composition of yeast and bacteria grown on normal paraffin. Yeast is less rich in protein: it contains only 50% instead of 68

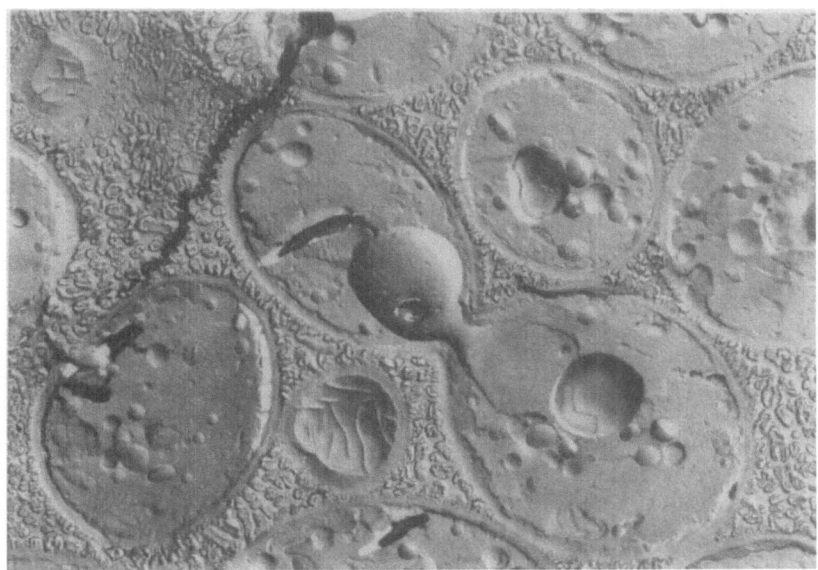

Fig. 3. Freeze-etching technique: Dividing cell

Table 5. *Yeast and bacteria grown on n-paraffins — chemical composition*

	Yeast %	Bacteria %
Proteins (N × 6.25)	50	60—72
Lipids	10	5—20
Ash	7	6—12
Sugar (by difference)	30	10
Humidity	2—4	2—4

to 72% in the bacteria. The lipid fraction in bacteria is rather important, but is highly dependent upon the nature of the different paraffinic cuts which are used in the feed-stuff.

Table 6 gives you an idea of the amino-acid distribution of the product; it is quite good, except for methionine, the value of which is rather low in yeast as well as in bacteria. In fact, our biological studies have shown that methionine is the onlylimiting factor despite the fact that, on a purely chemical basis, tryptophan looked also to be slightly limiting.

Table 7 gives you some biological data for bacteria and yeast: digestibility, biological value and net protein utilisation. This of course shows that the material is not an animal protein; but, anyway, it is satisfactory. What is interesting is the fact that the heat treatment in the case of bacteria improves the digestibility. The material behaves in this respect like vegetable protein.

Table 6. *Single cell protein — amino-acid distribution*

Amino acids (g/100 g)	Reference of F.A.O.	SCP from D_1 substrate	
		Yeast	Bacteria
Lysine	4.2	7.6	5.5
Threonine	2.8	5.0	4.1
Methionine	2.2	1.6/1.9[a]	2.0
Cystine	2.0	1.1	0.7
Valine	4.2	4.9	4.9
Isoleucine	4.2	5.4	4.1
Leucine	4.8	7.3	6.1
Phenylalanine	2.8	4.6	3.5
Tryptophan	1.4	0.8	1.0

[a] After performic acid oxidation.

Table 7. *Single cell protein — nutritive value*

	D	BV	NV or NPU
Bacteria			
dried	76.2	76.2	58.1
heat treated	88.4	73.6	65.1
Yeast	85	73	62

D = Digestibility
BV = Biological value
NV = Nutritive value
NPU = Net protein utilisation

Table 8 represents an assay of multiple amino-acid supplementation of the bacterial SCP. You can see that the protein efficiency ratio is in the neighbourhood of 2.6 when methionine is added, and that the addition of other amino acids does not improve the value of the material. As a reference, casein would quote 2.9 and the single cell protein without addition 2.1.

Here again, it would be of interest to find some adapted strains richer in methionine but, knowing that methionine is now readily available at low cost, it is certainly better, as a first step at least, to supplement the material with methionine.

Table 9 shows the fatty acid composition. It is dependent to a large extent upon the original feed. What is curious is that in those bacteria and

Table 8. *Limiting amino acid in bacteria*
Casein reference: PER = 2.9
SCP without addition: PER = 2.1

			PER
Methionine	Histidine	Valine	2.62
Threonine	Lysine	Isoleucine	2.12
Leucine	Tryptophan	Phenylalanine	2.08
2.58	2.00	1.89	

Table 9. *Fatty acid composition*

C_n	Fatty acid	(% on total fatty acids)		
		Bacteria on paraffin medium	Bacteria on non-paraffin medium	Yeast
C_8	caprylic	0.1	—	—
C_{10}	capric	2.65	—	—
$C_{10}=$	capro-oleic	5.25	—	—
C_{12}	lauric	2.95	6.10	—
C_{13}	tridecanoic	2.30	trace	—
C_{14}	myristic	14.95	2.05	—
C_{15}	pentadecanoic	13.50	trace	1.30
C_{16}	palmitic	19.60	37.35	15.20
$C_{16}=$	palmitoleic	2.60	24.80	14.00
$C_{16}{}^{br}$	phytosanoic (?)	—	—	0.80
C_{17}	heptanoic	12.90	—	—
C_{18}	stearic	3.10	2.50	2.00
$C_{18}=$	oleic	8.05	24.85	27.30
$C_{18}==$	linoleic	3.20	2.40	25.20
$C_{18}===$	linolenic	6.80	—	13.60
C_{20}	arachidic	2.05	—	—
Unsaponifiable (on total lipids)		35—50%[a]	4—7%[b]	5—7%[b]
Total fat (dry basis)		20—26%	6—10%	6—10%

[a] Mostly paraffins;   [b] mostly sterols.

yeasts fed on a paraffin medium, one finds odd number fatty acids, C 13, C 15, C 17, C 19 (C 19 is not quoted), which are generally not found in such amounts in biological materials. We have carried out nutritional

and toxicological studies on these odd number fatty acids, and they
appear to behave normally.

Fig. 4 illustrates the result of an interesting experiment. Pure
paraffin chains (hexadecan, pentadecan, etc.) were used as feed and the
fatty acid composition of the bacteria determined. Curiously enough,
each time one feeds an even numbered paraffin compound, one finds a
large amount of odd number fatty acids, which could be very easily
understood by the beta-oxidation at the end of the chain. In order to
avoid odd numbered fatty acid formation, we should use only pure
uneven numbered paraffins. This is, however, out of the question from
an industrial point of view.

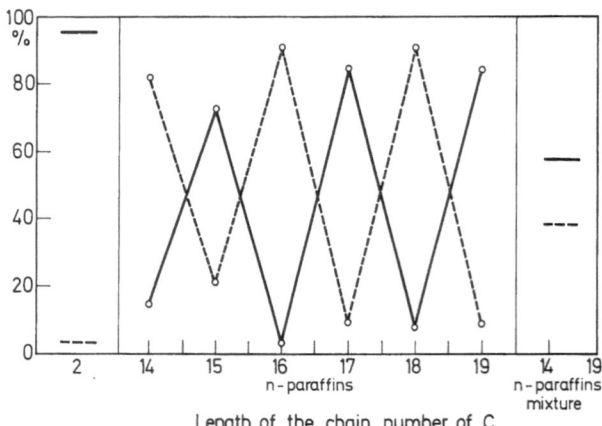

Fig. 4. Distribution of fatty acids in relation to the substrate. —— Total fatty
acids with even number carbon atoms. - - - - - Total fatty acids with odd number
carbon atoms

I have explained to you here, very briefly, one particular case of
Single Cell Protein development.

Before using this material as a food, we have to investigate it much
further and be sure that the material is wholesome, presents no toxicity
and can be used without difficulty.

First, we are carrying out *toxicological studies* looking for all the different aspects of toxicity, that means acute, sub-acute and chronic toxicity,
and the correlated problems like carcinogenesis and teratogenic power.
This is done in six different centers, and some of the investigations have
been carried out already over more than 3 years, using several species of
laboratory animals. We have been able to show that, by regular feeding
of this material, we have no impairment in growth, in general behaviour,
and in the main body functions of the animals. Fertility is not impaired
over several generations.

Besides these toxicological testings, we have been carrying out *immunological studies*. In fact, one of the difficulties is that we are dealing with a bacterial material and that you can be afraid that, even if the material is sterile, you can still have some remote danger of the presence of toxins or of an allergic reaction within the body from incorporation of this material. Our immunological research showed that the final product at the end of the whole technical processing has lost most of its immunological relationships with the live bacteria we start with. That means that the product has undergone on the way a heat denaturation, which obliterates most of the immunological properties of the native material.

In addition to these immunological and biological studies, we are planning *clinical testings* of this material, first on young adult volunteers, then on young healthy children and later on malnourished children.

What are *the prospects of the use of Single Cell Protein as a raw material for food*?

We are aware of the fact that the protein production in the world is decreasing every year relative to the increase in population, but we do not dream of one day replacing conventional proteins completely by Single Cell Protein. This will always be a new material which must be incorporated and mixed in complex food mixtures adapted to the nutritional requirements of the specific cases.

It is quite certain that in the future the use of hydrocarbon and hydrocarbon derivatives can yield a very interesting material for feeding human people.

British Petroleum's process is basically a refining process using biological means. The substrate is crude oil and the product yeast cells which must be cleaned and extracted to be used for feeding animals. In the Esso-Nestlé process which we are using now, we start from a substrate which is already of pharmaceutical grade. It is a paraffin cut which is extracted out of the refinery and which is processed in another place. And then, it will be transformed weight by weight into edible material for men; there is thus quite a difference in these two processes. This does not mean that the material which can be used for human feeding cannot be used also for animal feeding, but obviously there is a cost problem involved. Higher costs may be supported in the case of a human food, but they represent a major drawback when one is dealing with animal feed.

The Discussion

Rearranged by A. von Muralt

With 9 Figures

Contents

Introduction

It is an old experience that photographic images of a trip, projected in their chronological order, can be very boring. They are rendered attractive if one takes the trouble to rearrange them according to well chosen subjects. The same is true for reprinted discussions of scientific papers! (Some editors have even decided to refuse the publication of such discussions.)

I have tried to rearrange the discussions which took place after the presentation of the papers of this monograph in such a way that the ideas of the discussant were maintained in their original wording, but the sequence was changed, so that they fall into a pattern of different main subjects of interest.

The recording of every contribution by tape is a guarantee that nothing got lost, but I must bear the responsibility for having cut out all "small talk" which creates in a discussion a congenial atmosphere, but is without interest to those readers who have not joined the gathering.

A. General Mechanisms of the Body's Response to Protein-Calorie Malnutrition (PCM)

1. The Assessment of the Incidence of PCM

N. C. Wright opened the discussion and said:

I intend to deal only with one point, namely the inadequacy of available data regarding the incidence of morbidity and mortality due to PCM.

From the booklet entitled "Malnutrition and Disease" (WHO, 1963), published by WHO in connection with the "Freedom from Hunger Campaign", I would like to read three short passages. In the summary (p. 47), it is stated "Epidemiological studies, statistics of health, the results of experimental research and other sources of information on the relationship between malnutrition and disease reveal the importance of the problems of malnutrition and the particular vulnerability of the child under 5 years old. The most widespread form of malnutrition, protein-calorie deficiency, is a disease of childhood associated with weaning, the lack of suitable substitutes from internal lack and frequently failure to use available foods. The two main manifestations of this disease, kwashiorkor and marasmus, affect an unknown but vast number of children; not only do they often prove fatal in themselves, but they are contributory factors in the mortality attributed to many of the common communicable diseases."

This is a very excellent summary, except that it does not exactly reflect the evidence contained in the booklet itself. Clearly WHO realised this difficulty because earlier (p. 9) it says "unless a country possesses a good system of communications, a literate population, the means whereby sick people can be efficiently examined by medically qualified persons, and accurate records of the number and causes of disease and death, there will be little or no sound knowledge on which to base an idea of the extent of disease and the level of health. In few places do such conditions exist. Nevertheless, information from existing records, combined with results of intensive surveys and pilot studies, provide at least some indications of present conditions".

This again seems to me to be all to the good, but later, on p. 19, WHO gives an example of a survey, undertaken by INCAP, into the causes of death in children in a Guatemalan village. There the Civil Register undertaken locally indicated that, out of 109 deaths recorded, severe malnutrition (mostly kwashiorkor) accounted for only 1 case; but the INCAP's independent study indicated that the actual number was 40 out of 109 cases (practically 40%).

This does illustrate the great limitation in the amount of data we have. In order to check, I referred to FAO's (1963) Third World Food Survey, and was equally disappointed. There are a few individual figures given in relation to deficiency diseases, since generally the Survey deals with deficiency diseases *as a whole*. But it does conclude that about half of the world's population or between 1000 and 1500 million people suffer today from either undernutrition or malnutrition, or both.

There are two points which we should pursue:

a) I was very interested to see the details of the survey which is being undertaken by Dr. Martineaud in the Ivory Coast, in which he is classifying by a clinical study the population of a particular area. This is the sort of survey which should be undertaken if we are to get the facts and the figures on which we are to base our policies in the future, whether in individual countries or on a world-wide basis.

b) The second point which I should like to make is that the papers of this symposium, though they may appear somewhat academic in their outlook, should in fact help us towards finding a far better index of the incidence of such malnutrition. In regard to PCM, we are only seeing at present the tip of the iceberg; the greater part is below the surface. It appears to me that extensive biochemical and other clinical tests such as those discussed today should go far to fill this gap in our knowledge.

2. Food Requirements and Quality of Food

J. Mauron:

Is there not a possibility that dietary allowances are influenced by the food consumption in wealthy countries and that these allowances are then considered to be valid and even compulsory for mankind in general ?

A. E. Harper:

I think there are misconceptions about the dietary allowances as formulated by the National Research Council of the USA. Historically, these were devised for mass feeding with the objective that, if the food supply contained the amounts of nutrients recommended, there would be little likelihood of any nutritional deficiency in the entire population. They are not *minimal requirements* for human subjects. One of the main uses for them is in institutional feeding or the feeding of large groups, such as the armed forces, where they serve as a guide for the procurement of a nutritionally adequate supply of foodstuffs.

R. G. Whitehead:

I think one of the reasons why in many tables there are such high values for protein is that it is very difficult to devise a diet using European foodstuffs which contains enough calories without having much more than the minimum amount of protein needed. We eat maybe 2500 Calories each day and, if we do this with our normal food, then it is inevitable that we eat 70 or 80 g of protein.

C. Gopalan:

I think it is true that the dietary allowances which are currently recommended were largely determined by the consumption patterns prevalent in some better-nourished countries of the world. However, during the last few years, the FAO and the WHO have brought together groups of people who have attempted a downward revision of these dietary allowances. Generally speaking, in earlier years, we were inclined to exaggerate the dietary requirements. In this connection, it is very necessary to make a clear distinction between "requirements" and "allowances".

G. Arroyave:

At INCAP, we try to establish a figure of the protein requirements on an experimental basis. Protein of egg, between 1 and 1,25 g per kilo of body weight, is sufficient to reverse all the initial changes produced by a low protein diet, such as a reversal of the non-essential/essential amino-acid ratio, and a shift in the hydroxyproline excretion to the higher level.

7*

3. Homeostatic Factors regulating the Protein Metabolism

3.1. Neural Factors

H. Isliker:

I would like to ask what is known with regard to the question which Dr. von Muralt raised this morning, namely the mechanisms which are responsible for the homeostatic regulation of the protein matrix. What are the signals which in the case of protein deficiency inform the body to "sacrifice" liver and tissue proteins in an inverse order to their relative importance? What are the centers which are responsible for this homeostasis?

A. von Muralt:

There is a difference between the neurophysiologist's approach and the biochemist's work. Dr. Harper, did you implant in your animal-experiments permanent electrodes into the *hypothalamus*? One in the central region and one in the lateral region which are antagonistic? What sort of activity did you get if a decrease in appetite appeared if you give an amino acid-deficient diet to your animals? Could you compensate this by an artificial stimulation of the feeding centers? I could imagine that a deficiency in certain amino acids affects the very sensitive feeding center, so that its activity goes down, which might be the reason why you found a decrease in the weight of your animals.

I have the impression that those scientists who are working on the neurophysiological side are using crude methods in nutrition, and those who are working on nutrition forget that there is the possibility to implant electrodes in their animals and to lead off the electrical activity of the hypothalamus as Anand (1961) has done. We should encourage every cooperative effort between those who are working on central regulation and those who are interested in the biochemical aspect of nutrition.

A. E. Harper:

Some preliminary results were obtained with electrodes implanted in the *feeding center*, indicating that it took a greater stimulus to get the animals to eat an imbalanced diet.

There have been three studies on animals with lesions in the *satiety center* of the hypothalamus. In one study, food intake of lesioned animals fed an imbalanced diet was not depressed (Nasset et al., 1967). In this study, the animals were somewhat older and their protein requirement was probably quite low. Under such circumstances, they are usually less sensitive to an imbalance.

In two other studies, one by my former colleagues Rogers and Leung (unpublished), of the University of California, Davis (1968), and another by Krauss and Mayer (1965), lesioned rats that were fed the imbalanced diet responded just like intact rats.

R. G. Whitehead:

I am glad the hypothalamus has entered into our discussion. I am quite sure that Dr. von Muralt was absolutely correct in introducing the term homeostasis, and the importance of homeostasis in our consideration of nutrition.

There must be, for example, a homeostatic mechanism responsible for the maintenance of amino-acid concentrations in the serum, because these patterns are remarkably constant in man, chickens, rats, pigs, cows, horses and baboons. As far as I know, nobody has really studied the homeostatic mechanism controlling this.

3.2. Endocrine Factors

G. Arroyave:

The *endocrine system* may play an important role in homeostasis and, as a consequence, in the development of either the kwashiorkor type of malnutrition in children or the marasmus type.

Kwashiorkor is described from the etiological point of view as a malnutrition type which develops when a child is practically force-fed a diet rich in carbohydrates and poor in protein, both in quality and quantity of protein.

And marasmus develops when a child at a rapid growth age is insufficient in both calories and proteins, a type of starvation diet.

Cortisol or hydrocortisone, or cortisone in a rat will provoke or promote protein synthesis in the liver and will produce protein catabolism in the muscle. The pancreas producing insulin will act in the muscle, producing or stimulating the incorporation of amino acids intracellularly.

Now, if we think of a child which has been submitted to a force-fed diet rich in carbohydrates, we may think of an inhibition of the production of glucocorticoids because there is no need for gluconeogenesis. In this case, the action on muscle wasting is weak, and there would be a tendency to concentrate amino acids in the muscle mainly because the diet, rich in carbohydrates, acts on the pancreas to produce more insulin to keep the glycemia down. This provokes in turn a concentration of amino acids intracellularly in the muscle, and a decrease in the concentration of amino acids in the hepatic pool.

Under these conditions, we would have a true amino-acid deficiency in the liver because no amino acids are derived from the diet in sufficient amounts, and no amino acids are coming from the muscle to the liver

because of the impaired mechanism. This would be a conditioned imbalance in endocrine function provoked by the force-feeding of an excess
of carbohydrates. This agrees with the clinical picture of kwashiorkor
which under certain circumstances has been called the "sugar baby", a
child who still has quite a significant reserve of muscle tissue. He has a
liver with a low concentration of proteins and full of fat, and he is protein
malnourished.

The marasmic type is a child who has had, for a long time, a very low
supply of calories in general. So the marasmic child has to develop or
stimulate the mechanism to catabolise his own proteins in order to supply
the energy necessary for survival. Under these conditions, the adrenal
would be producing a lot of glucocorticoids to act upon the muscle, to
catabolise the muscle and to make the amino acids available for protein
synthesis at more important sites and for more important proteins.

The anatomic pathology of a marasmic child fits very well into this
picture: muscle progressively exhausted to the zero limit; but the liver
of a marasmic child is a red liver without fat, functioning well, and with a
high concentration of protein per gram of tissue. Due to the stimulation and good performance of the endocrine system, the marasmic child
is in the fortunate position of being able to eat himself, and he will die
when he finishes this supply of food which is the muscle.

C. Gopalan:

This is a very attractive hypothesis; we have also been thinking on
these lines. The plasma cortisol levels are no doubt raised in marasmus
to a significant extent, but they are also raised in kwashiorkor. The only
difference is that response to synacthen is exaggerated in the marasmic
state. This hypothesis does not explain why cortisol levels are elevated in
kwashiorkor.

G. Arroyave:

Our observation, the one which incited us to elaborate this
hypothesis, showed that the excretion of hydroxycorticoids in the kwashiorkor child was much smaller than in the marasmic child. We did some
rat experiments in which the rats were put on a low-protein diet (5%
corn protein); to one group we gave cortisone and to the other we did not.

The one with the poor protein diet plus cortisone was thrown into a
very typical marasmic state, much more rapidly than the other; they
died or fell into inanition with practically no muscle, with a very red and
good-looking liver. The other got the symptoms characteristic of a rat
on a 5% corn protein diet: fatty liver, and impaired growth. The excretion of the steroids in the urine is a very poor approach and the levels of
cortisol in the blood also, because it depends on the amount of bound

and free cortisol, and on many other circumstances. The only way to solve the problem would be to determine actual secretion rates of gluco-corticoids by the adrenal. This has to be done with labelled cortisols, and it is extremely difficult to do it in children. But that would provide the only answer as to the functioning capacity of the adrenal cortex under the circumstances.

C. G. King:

Could this be done in vitro by taking the tissue from children who have died, just as they do for transplanting ?

G. Arroyave:

I had not thought of that. There are many other hormones such as growth hormone and others that I did not mention because I do not know how they would fit here.

S. Frenk:

Formerly, when adrenocortical function was mainly assessed by the measurement of steroid excretion in urine, one could argue that, since production of a fatty liver requires appropriate adrenal activity, the interesting findings of Castellanos and Arroyave (1961) on malnourished children could perhaps be explained by a decreased glomerular filtration rate, causing low hormone excretion in urine. Now, measurements of plasma cortisol by Alleyne and Young (1967) have actually shown high levels in severely malnourished children, in the same way as Pimstone, Wittman, Hansen and Murray (1966) had observed raised fasting plasma levels of growth hormone in similar patients. Moreover, plasma cortisol concentrations show no circadian rhythm and are only partially suppressed by dexamethasone, in the same way plasma growth hormone levels may fail to react to the suppressive effect of glucose. These findings suggest that there is a peripheral or central block to the regulatory actions of these hormones in severely malnourished children.

Just from the purely endocrinological standpoint, it is not acceptable, if the general mechanisms for maintenance of hormone levels and activity are operative in malnourished children, that they have a high level all the time. Functional studies done by several scientists have shown that there is no possibility of depressing these high levels. You can give hexamethasone and the cortisol levels stay high; you can give glucose and the growth hormone stays high; and in normal cases, it is well known that if you give glucose, for instance, you will depress your levels of growth hormone.

All this, I think, points or at least suggests that, in these kinds of patients, there is a block for the action of these hormones, both peripheral and in the central nervous system, wherever this control is. In the presence of high plasma levels of cortisol, you have low urinary levels of hydroxysteroid; this is just another way of demonstrating that these children have a decreased glomerular filtration rate and that therefore we cannot rely on anything which is measured in urine in these children.

A. von Muralt:

Have you ever taken into account that possibly a protein deficiency could change the sensitivity of the receptor site for hormones in the tissues so that a high hormone level does not mean necessarily that there is a higher activity of the hormone, but that the receptor site became less sensitive to that hormone ? The level of the hormone is therefore raised as a compensation.

S. Frenk:

That certainly is a good possibility.

3.3. Metabolic Pathways

R. G. Whitehead:

I would like to express Dr. Arroyave's ideas in a biochemical form (Table 1).

I am going to suggest why a distortion in the serum amino acids might develop in a situation when the child is being fed on low-protein, high-carbohydrate diet, but not when the child is being fed on a low-protein, low-carbohydrate diet. The human body is pretty well adapted, as far as homeostasis is concerned, to deal with starvation.

Table 1. *Metabolic cycle*

Probably this is a normal situation for, in the natural state, especially in hunting tribes, man is an intermittent eater, so he has to be physiologically adapted to withstand this stress to his homeostasis. He can of course utilise body fat or protein to provide energy, as shown in the diagram. But what happens if his diet has adequate amounts of total calories but too little protein?

There is less tendency under these circumstances to break down protein or fat, because there is plenty of energy in the diet.

Thus there is an essential difference between primary protein deficiency and calorie undernutrition. In the latter, there must be a steady supply of the essential amino acids from the tissue proteins, should they be needed, whilst in the former state this probably does not occur to anything like the same extent. This could be why a distortion in the serum amino-acid pattern occurs in protein deficiency, but not in marasmus.

3.4. Muscle Wasting

D. B. Jelliffe:

What do you feel is the mechanism which leads to the very severe *muscle wasting* in kwashiorkor? How does that fit into your theory?

G. Arroyave:

I understand that, in the case of "sugar baby", there is no particular muscle wasting.

D. B. Jelliffe:

Well, this was a point which had occurred to me when you were describing your two extremes: marasmus and the other extreme: kwashiorkor.

The sugar-baby type kwashiorkor to which you are referring does not seem to be seen so much nowadays, although it was common when I worked in the West Indies some 15 years ago.

These cases on the whole were quite noticeably less severely ill. They could stand up in the cot and had little anorexia. When handed a mug of milk, they would drink it and would get better without elaborate treatment. They had a low mortality, therefore the autopsy findings on this type of sugar baby were scanty.

There is a clinical impression that the child with sugar-baby kwashiorkor probably had only modest muscle wasting. But my question does not refer to this group which is, after all, rather an extreme variant of the kwashiorkor syndrome.

I would like to ask why in ordinary kwashiorkor there is severe muscle wasting.

G. Arroyave:

When a child gets to the hospital, we look at him, we do some tests and we classify him either as marasmus or kwashiorkor. What do we know of his history in the past 6 months to 1 year?

When we have been able at INCAP to obtain clinical histories in the past 6 months, we have obtained sometimes curves which are very, very interesting. It is a child who developed oedema 6 months before. He was manipulated at the town or the village or by the witch-doctor, and the oedema disappeared; then he developed oedema again, and the oedema disappeared; then the third time and then the fourth time; then he gets very seriously ill and gets to the hospital. So what is he? From the historical point of view, he is a good mixture of anything. He has had kwashiorkor and has marasmus; he has had diarrhoea, extreme fall in inanition because, once they get diarrhoea in the field, they are put on sugar waters or on very low diets. So he is only a consequence of all these episodes that provoke him to have muscle wasting, oedema and everything else at the same time.

D. B. Jelliffe:

May I ask a supplementary? In my personal classification of kwashiorkor, I regard muscle wasting as one of the cardinal features. Is muscle wasting a characteristic symptom of kwashiorkor and, if so, how is this brought about biochemically?

A. E. Harper:

In animals fed a diet with a rather mild amino-acid imbalance, we were able to prevent the food intake and growth depressions by injections of insulin. We attributed this to the influence of insulin on blood glucose utilisation, according to the glucostatic theory of food intake regulation overwhelming any depressing effect from the surplus of amino acids. Unfortunately, animals fed the diets with severe imbalances died from hypoglycemia before they would eat.

However, when the animals fed the diets with more severe imbalances were treated with cortisol, which classically depresses the growth and food intake of the rat, although control animals showed the classical depressions, food intake and growth rate of those fed the diets with amino-acid imbalances were enhanced. Thus both growth and food intake of rats having an abnormal blood amino-acid pattern were stimulated rather than depressed by injections of cortisone.

C. Gopalan:

Those who have had opportunities to study poor communities over prolonged periods will realise the fact that marasmus and kwashiorkor are not just two completely different and etiologically unrelated syndromes.

In communities subsisting on the same deficient diet, marasmus and kwashiorkor coexist. They not only coexist in the same community, they coexist in the same individual at different points of time.

We quite often see a child who is marasmic and after a couple of months becomes a subject of kwashiorkor. Again after some time, when the oedema clears off, the child presents the picture of marasmus. When a child who was previously marasmic develops kwashiorkor and shows muscle wasting, we designate it as "marasmic-kwashiorkor", adding to the confusion. Marasmus and kwashiorkor are but two clinical facets of protein-calorie deficiency, and must not be considered as two distinct and different disease states with varying etiology.

As Dr. von Muralt very rightly pointed out, there is a relative insensitivity of the target organs to these hormones, and it is quite likely that a very high concentration of this hormone has to be maintained. It is possible that at some stage there may be an exhaustion of the adrenal, and the levels come down to a point not sufficient to produce the mobilisation of the muscle proteins and their deviation to the liver.

D. B. Jelliffe:

I agree with Dr. Gopalan and Dr. Arroyave about the potential convertibility of one severe syndrome to another, but I do feel that many cases can be put in water-tight compartments. For example, in the Caribbean you can see cases of kwashiorkor which occur at the age of 3 months and where you know the history of the child all the way along; these are cases of kwashiorkor ab initio if you like. They have not gone through any marasmic phase and they do have wasted muscles.

C. Gopalan:

I just want to clarify this. I do not deny that classical kwashiorkor does occur in certain situations and classical marasmus may be seen in young infants who are getting very small quantities of breast milk. But when we are talking of marasmus over large parts of the world, we are talking of the child of the same age-group as that of kwashiorkor. In this type of marasmus, the underlying dietary situation is not different from that obtaining in kwashiorkor.

R. G. Whitehead:

Could we *look* at two children, one with kwashiorkor (Fig. 1), and the other with marasmus (Fig. 2).

D. B. Jelliffe:

The child shown with kwashiorkor has got practically no trapezius and no pectoralis major muscles. The arm is extremely wasted. If you were able to palpate this child's arm, you would feel only a small mass of wasted muscle.

The marasmic child has a much greater degree of wasting of muscle. This I completely agree with but, at the same time, it is "unmasked", because there is no fat and no oedema. In kwashiorkor during therapy, as the oedema slides off, the muscle wasting becomes more obvious.

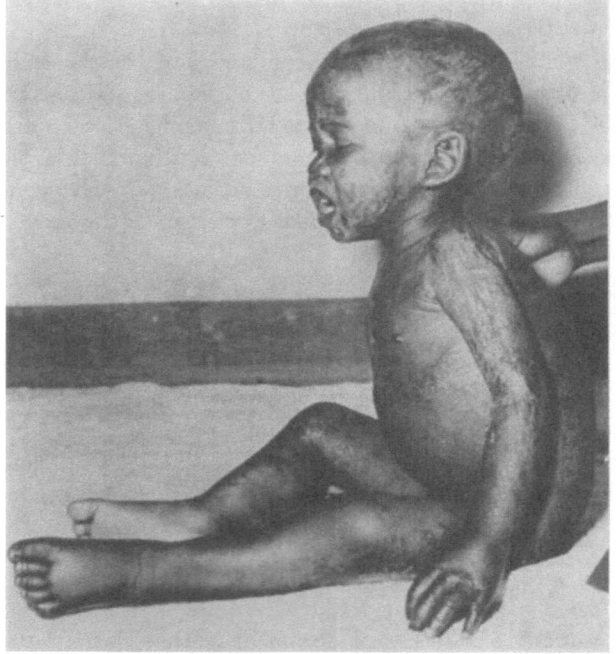

Fig. 1. Child with kwashiorkor

A. von Muralt:

What is the primary cause of muscle wasting in polio?

We should not forget that there is an important neural factor. Let us assume that the brain is very sensitive to protein deficiency. Then there might be an additional influence coming from the central nervous system inducing muscle wasting through a decrease of the tonic innervation of the muscle.

G. Fanconi:

This question is very interesting. We believe from our experience in poliomyelitis that a center in the diencephalon is involved in the calcium regulation. If this center is involved in some cases of poliomyelitis, calcium may be lost into the urine and decalcification of the bones is the consequence. There might be something like this with regard to the tonic innervation of the muscle.

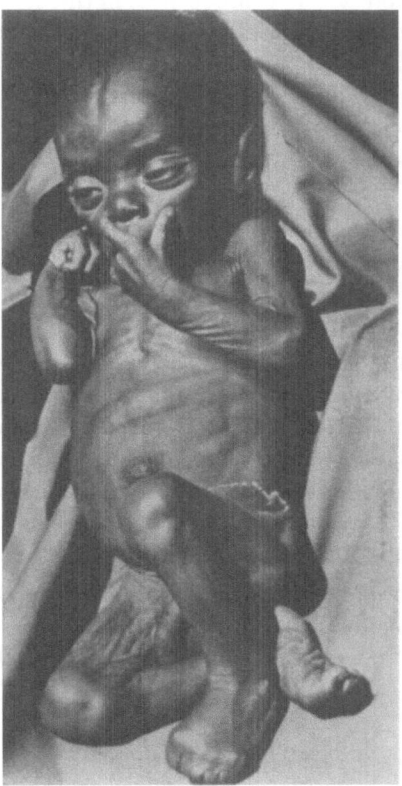

Fig. 2. Child with marasmus

A. von Muralt:

The lack of tonic innervation produces the wasting due to a change of the enzyme pattern of the muscle. Dysuse precipitates this development. It is possible that hypo-protein supply to the brain diminishes the activity of those centers which are responsible for the tonic innervation of the muscles.

3.5. Homeostasis — Homeorhesis

D. Bovet:

There is a problem in what Dr. von Muralt called the problem of *homeostasis*. He showed the mechanisms of regulation during hunger. I think the same concept of homeostasis was developed by Waddington (1957) who called it *homeorhesis*, to show the dynamic aspect of the problem.

This aspect appears in two cases: Dr. Whitehead showed that a low-protein diet has more severe consequences in the presence of an excess of carbohydrate. Second example: the case of phenylketonuria in which the presence of one amino acid provokes the formation of a metabolite which is responsible for very severe injuries to the developing brain.

So, as Dr. Fanconi says, nutrition is not only a problem of quantity. In the future, we shall have to know more about the quality of the food.

B. The Appraisal of PCM

Introduction

Biochemistry has made such spectacular advances with regard to knowledge, techniques and apparatus that a new and promising approach exists as to the appraisal of PCM, not only in its clinical stages, but also in the adaptive and subclinical stage, where the syndrome only begins to manifest itself. This statement needs an explanation.

A constant deficiency of protein in the diet gradually evokes reactions of the body, which go through the following stages:

1. Range of physiological adaptation to a hypo-protein diet.

Homeostatic factors come into play, reduce the metabolism of proteins and maintain the nitrogen balance at a low level of excretion. Result: change in the enzyme-pattern and the hormonal control in the body. Muscle wasting.

2. Failure of these adaptive mechanisms.

The protein metabolism becomes increasingly unbalanced. Subclinical symptoms are noticeable and many biochemical changes are apparent.

3. Serious damage to the organism.

Clinical symptoms, such as skin lesions, oedema, fatty liver, discoloring of the hair, and mental apathy appear. The protein metabolism is seriously unbalanced and the biochemical equilibrium disturbed. *Malnutrition has produced malfunction in the body.*

The precision of modern biochemical methods is such that it is now possible to study not only phase 3, as has been the case till now, but also phase 2 and even phase 1! There is no single biochemical assay which will give enough satisfaction in such studies, but we feel that the assembly of what might be called "a biochemical battery" may give interesting and significant results. A large part of the discussion centered around this problem. At the end, it was decided to make a well coordinated effort between the group of Dr. Whitehead (Kampala, Uganda), the group of Dr. Arroyave (INCAP, Guatemala) and the group of the *Nestlé Foundation* in Adiopodoumé (Ivory Coast), by using the same methods, the same equipment and the same approach to the problem. An exchange among the members of the three groups, regular working meetings and a flow of information regarding methodological improvements were considered to be essential for success.

1. Biochemical Tests

1.1. Level of Plasma Proteins

G. Fanconi:

Dr. Bengoa, you mentioned the decrease of protein in the plasma as a sign of malnutrition. If children with nephritis were fed a very low-protein diet, the protein in their plasma decreased, but these children were not at all ill.

My question is: What is the degree of malnutrition if the protein level decreases in the blood?

J. M. Bengoa:

I think Dr. Fanconi is right in saying that the *protein in plasma is not really a good indicator of malnutrition*. As a matter of fact, we have not taken the total protein as an indicator of protein deficiency. In most of these surveys, which have been undertaken not only by WHO but by many other institutions, the albumin level is a better indicator.

A. von Muralt:

Within the range of physiological adaptation to a hypo-protein diet, the homeostatic reactions maintain the levels of the plasma proteins within the physiological range. A marked drop is therefore a sign of *failure of the homeostatic equilibrium*.

1.2. Excretion of Protein Metabolites

R. G. Whitehead:

Apart from serum protein measurements, the biochemical test which probably is in most general use is the *urea/creatinine ratio* (Dugdale and

Edkins, 1964). This ratio reflects the *dietary status* but not the *nutritional status*. This is important to understand. As soon as a person goes on to a low-protein diet, the amount of urea excreted relative to the amount of creatinine naturally falls. This does not mean to say that the urea/creatinine ratio is not an important measurement. It can be of value in connection with dietary surveys. It is very difficult to do accurate dietary surveys, so extra information is always welcome!

The next test, which is one of my own, based upon distortions in the serum amino acids, is the *amino-acid ratio* (Whitehead, 1964). This is a semi-quantitative method, based on paper chromatography, introduced to cut down the high cost of more elaborate estimations, and an attempt to design a method which could be run by semi-skilled and semi-trained technicians.

If a child is becoming malnourished because of a low-protein/high-carbohydrate containing diet, the amino-acid ratio becomes elevated because the serum amino-acid pattern is distorted. I emphasise, this test is limited to areas in which the children are being made ill by eating a *low-protein/high-carbohydrate containing diet*. The area around Kampala, in Uganda, is a good example of this.

In another part of Uganda, however, called Karamoja, in which there is a general shortage of food, quite normal amino-acid ratios are found.

So the important thing to understand is that the value of this test depends upon the type of area being investigated.

The third test, which was again proposed by myself, is based upon the *excretion of hydroxyproline-containing peptides* in the urine. The advantage of urine analysis is that you can obtain urine without undue trauma to the child. This is important if you are going to do surveys sitting under a tree in the "bush", and you depend on the cooperation of the mothers! The disadvantage of urine is that you cannot collect a 24 h sample very easily under these conditions.

Originally, the hydroxyproline excretion was expressed as the ratio between hydroxyproline and creatinine. Between the ages of 0 and 5 years, this ratio fell gradually in quite normal children, and I did not want a ratio which changed with age, so I put in a weight factor to produce a more constant index. The hydroxyproline index was described in *The Lancet* (Whitehead, 1965).

The theory behind this test is that, when growth is slow, collagen synthesis in the connective tissue is reduced, and small amounts of hydroxyproline are excreted in the urine.

Thus, hydroxyproline excretion is related to growth and a reduced rate of growth is found in all forms of protein-calorie malnutrition.

The main disadvantage of this test is that hydroxyproline excretion is sometimes elevated in severe kwashiorkor, presumably because of a raised breakdown of structural collagen or possibly the tropo-collagen precursors of collagen. As far as I know, this difficulty has only been reported from Uganda.

J. S. Dinning:

Is it low with low growth or high with low growth?

R. G. Whitehead:

The excretion of hydroxyproline peptide is very low when the rate of growth is low. This is true of hormonal dwarfism as well as nutritional dwarfism. It reflects collagen metabolism which is indirectly related to growth.

The fourth proposed test is the *creatinine/height index* (Viteri, Arroyave, Behar, 1966). This test is based on the fact that malnourished children have less muscle than normal children; and although creatinine excretion per unit weight (creatinine per kg) is not so reduced in malnutrition, creatinine per height is. The snag with this test, for field use, is that timed samples (12 or 24 h) of urine need to be collected.

What about the future? Dr. Arroyave and I, as you can see, are both working towards a *battery of tests*, or a metabolic profile.

We want tests which reflect metabolic abnormalities in different tissues: in connective tissue, in the liver, in muscle, etc.; this is what we are aiming for. And of course, the separation of isoenzymes may very well be important in this respect.

But may I make a plea? *On biochemical grounds, we should not deem a child malnourished unless the biochemical test has previously been shown to be related to malfunction.* In other words, malnutrition should not be diagnosed merely because a child has had to modify his metabolism to maintain homeostasis. This is not malnutrition; the child has just responded to the changed diet.

A battery of tests implies automation and multiple analysis. In clinical biochemistry, we are already using these techniques for diagnostic purposes. It is much more easy to interpret biochemical data if we have other metabolic information with which to compare the more specific test. There is of course the problem of cost, but it has been shown that, once you introduce automation, the cost per analysis falls; the large cost is in the initial outlay.

We have, of course, yet to prove our case, but I think honestly that the future is with biochemistry in the diagnosis of pre-kwashiorkor.

D. B. Jelliffe:

Is there any biochemical test at the moment which can in any way replace the weight for age as a simple method of assessing PCM in the field?

R. G. Whitehead:

There is no single biochemical test which can replace weight for age, but I think we are getting now to be able to provide a suitable "metabolic profile".

D. B. Jelliffe:

Biochemical tests will undoubtedly play a useful part but, under the practical circumstances, one has to think in terms of simplicity.

N. C. Wright:

I would just like to ask one question to Dr. Jelliffe: If you were provided with sufficient funds for automated equipment, would you use this equipment?

D. B. Jelliffe:

If I were supplied with an automated laboratory, I most certainly would accept it and use it with a great deal of pleasure. The reason I would do so would be because I would like to hope that the different tests would give information concerning the functions of various body tissues which could be correlated with simple anthropometric measurements.

1.3. Activity Level of Enzymes

1.3.1. Xanthine Oxidase

J. S. Dinning:

One of the enzymes in experimental animals which is most sensitive to protein deficiency is *xanthine oxidase* in liver.

Xanthine oxidase is present in the serum. Have there been any studies of serum xanthine oxidase in protein-calorie malnutrition?

This question found no answer and remains open! (Ed.)

1.3.2. Lysosome Enzymes

G. Fanconi:

Dr. Aebi, you mentioned that lysosome enzymes are increased in malnourished children. They are increased, this is my belief, because lysosomes are the suicide-bag in the cells. If the child is malnourished, many cells will be destroyed, so the lysosome enzymes will increase.

J. S. Dinning:

I was particularly interested in one of the figures of Dr. Aebi on the excretion of arylsulfatase by Indian children with kwashiorkor (the elevated excretion). It was pointed out that this is *lysosomal enzyme*. In vitamin-E-deficient animals, the most striking biochemical change is a great increase in the concentration of lysosomal enzymes due to rupture of the lysosomes. Patients with kwashiorkor frequently exhibit low levels of serum vitamin E, and it has been reported that, in some instances, from two different locations the anemia accompanying kwashiorkor will respond to vitamin E.

It would therefore be interesting to treat these children with vitamin E and see what it would do to the excretion of the enzyme in the urine.

H. Aebi:

The experiments with arylsulfatase indicate that an increased amount of arylsulfatase A is excreted in urine. This seems to be a very specific phenomenon since there was no significant increase in the excretion of an isodynamic enzyme, the arylsulfatase B.

The fact that an increased amount of urinary excretion can be observed is not necessarily due to the fact that synthesis or accumulation is increased. It merely reflects that the permeability barrier of the suicide-bag is altered. Vitamin-E deficiency may be the most important factor for this increase in urinary arylsulfatase excretion.

1.3.3. Lactase

G. Fanconi:

Dr. Aebi mentioned that the change in diet modifies the effect of enzymes. I was very impressed that children with galactosemia who did not receive any lactose during their life showed a lactase activity in the duodenum, which is completely normal. How to explain this, if you mean that a change in the diet can modify the enzyme activities?

H. Aebi:

Your question as to the enzyme pattern in the intestinal mucosa in cases of galactosemia: It is possible that some enzymes do not show any change; however, many of them do. Often the activity of a group of enzymes is altered in a constant proportion, sometimes changes are erratic. Therefore we should have a look at all possible enzymes, so that we can select the right ones.

R. G. Whitehead:

I quite agree with Dr. Aebi that the study of enzyme patterns offers theoretically *very exciting possibilities for assessing nutritional status*, but I wonder whether I could give a warning as to what might happen if these were used for the assessment of malnutrition.

Enzyme patterns might fluctuate wildly and reflect recent dietary inadequacies, because they are an important part of the homeostatic mechanism.

If we wanted to use enzyme patterns as a measure of nutritional status, we would have to interpret them rather carefully.

G. Fanconi:

Dr. Whitehead insisted that lactase deficiency is a consequence of genetic polymorphism. I do not believe that he is completely right. Durant (1958) described the idiopathic lactosuria; I said in the discussion that the lactosuria could be secondary because in malnutrition the intestine gets permeable to lactose. Last year, Durant agreed that probably the lactase insufficiency is not only a genetic defect, but also a consequence of malnutrition. In Zurich, Shmerling (1964) could demonstrate that, in every case of coeliac disease and of kwashiorkor in small children and infants, we have to do with a secondary decrease of lactase in the intestine. The recuperation of lactase goes much more slowly than the recuperation of the other saccharases. We believe that the lactase insufficiency is generally a secondary one and not a genetic one.

R. G. Whitehead:

I would like to thank Dr. Fanconi for making these comments, and I hope I did not sound as if I were insisting our findings were due to a genetic difference. We were not completely convinced that it was genetic, but this was a possibility. The children never recovered their lactase activity even with first-class treatment. The Baganda doctors who were well-fed, often coming from rich families, also had a reduced lactose tolerance.

1.3.4. Enzyme-Protein Synthesis

H. Isliker:

We produce each day several liters of intestinal fluid, representing an exceedingly large amount of hydrolytic enzymes. As one knows, the synthesis of these enzymes is located essentially in the intestinal mucosa which is able to utilise absorbed amino acids as starting material. I wonder to what extent a reduction in protein intake may be responsible for a decreased enzyme synthesis in the intestine, which in turn may reduce the absorption of proteins. The existence of such a vicious circle could be studied in well-fed and starved animals by a variety of techniques.

A. E. Harper:

I do not have an answer to this. The production of protein and the turnover of protein in the gastro-intestinal tract, which can be very substantial and even exceed the protein intake of the organism, represent essentially re-cycling and continuous re-utilisation so that "what the gut giveth, the gut taketh away". It should, therefore, have little influence on protein requirements unless digestive enzymes are so depleted that protein digestion becomes inefficient — a rare condition in so far as I know.

1.3.5. Alkaline Phosphatase

S. Frenk:

Serum alkaline phosphatase is decreased characteristically in protein-calorie malnutrition, at the expense of the bone isoenzyme. When rickets is superimposed, a frequent phenomenon in our city, the high alkaline phosphatase due to the latter appears to be cancelled by the decrease induced by overall nutritional failure, resulting in apparently normal serum levels of this enzyme.

H. Aebi:

In clinical chemistry, two types of information are collected when measuring the activity levels of enzymes:

1. Direct evidence is obtained only by looking at the tissue itself with biopsy, or at white or red cell enzymes. This is a field which has not yet been exploited in detail.

2. Indirect evidence can be obtained much more easily; however, the enzyme levels in serum or in urine are not necessarily directly correlated to what happens within the cell. My specific questions then are:

What are your experiences in getting biopsies?

What is known about N-loss by feces in malnutrition?

1.4. Biopsy as a Means of obtaining Biochemical Data

D. B. Jelliffe:

I would like to just mention a word here on biopsies. Of course, there are two types of biopsies: those that you can do in hospital, and those that can be obtained in the field.

The hospital biopsies that have been done in PCM include liver — Waterlow's well-known work on liver enzymes —, and also muscle biopsies — again principally by John Waterlow and his co-workers.

Problems of obtaining *biopsy tissue in the field* are plainly much greater. Liver biopsies have been done in the field in the past, but I personally would regard this as unwise.

There are, however, two types of field techniques that are practical; one is taking a smear of buccal mucosa cells, which is, in a sense, a biopsy of living material. The other is a technique which Dr. R. Bradfield of California has introduced, whereby hair root morphology is examined. Hair is relatively accessible and can easily be plucked out; the morphology of the hair roots can be examined later at leisure. Hair is after all a protein tissue and the root is a rapidly growing zone. Dr. Bradfield has classified these changes in hair root morphology, and in severe protein-calorie malnutrition there are very striking differences compared with the normal.

C. Gopalan:

It may not be possible to obtain liver biopsy specimens in field studies. We investigated the possible use of estimation of cysteine content of hair as an index of protein-calorie malnutrition and we have shown that this is decreased in protein malnutrition.

H. Isliker:

I would like to draw attention to the determination of albumin in body fluids other than blood. There is about twice as much extravascular albumin as that present in the intravascular pool and certain authors like Hoffenberg et al. (1962) have observed a preferential loss of extravascular albumin in protein deficiency.

In the case of saliva, a most easily accessible body liquid, the concentration of albumin is small but fairly constant in normal individuals. A few years ago, a new method of immunodiffusion became available (Mancini, Carbonara, Heremans, 1965), which allows a precise determination of albumin in quite small concentrations to be carried out. Using this technique our laboratory, jointly with WHO, has carried out field tests in Africa for the determination of γM-globulins. The technique is simple and reproducible, and requires only trace amounts of samples. It could be used as a preliminary screening procedure.

D. B. Jelliffe:

One should certainly look into these rather unorthodox approaches to studies of unconventional tissues such as tears, the ear cartilage or the physical and chemical properties of the hair. But the saliva is problematic because the composition from different glands differs. However, there is a simple sucker-like device with which one can obtain saliva from the parotid gland alone.

G. Fanconi:

We have some experience with saliva, because in cystic fibrosis the increased level of NaCl in the saliva is one of the most significant symptoms of the disease. We know today that the composition of the saliva depends on the drinking habits and on the age of the person. It is very difficult to draw conclusions from the composition of saliva.

S. Frenk:

Returning to *urine* as a specimen for study, due emphasis must be put on the fact that, regardless of the clinical type, children hospitalised with severe malnutrition show a reduced glomerular filtration rate (Kerpel-Fronius, Varga, Kun and Vonoczky, 1954; Gordillo, Soto, Metcoff et al., 1957; Alleyne and Young, 1967). For a relatively easy appraisal of glomerular filtration rates under field conditions, the use of radioactive compounds such as the chromium-EDTA-complex-^{51}Cr may be taken into account. In strict terms, indices of nutritional status based on measurements made on urine ought to be corrected for the status of renal function. The performance of percutaneous *skeletal muscle biopsies* for assessment of nutritional status under field conditions has become a real possibility with the use of the device recently described by Nichols, Hazlewood and Barnes (1968).

H. Aebi (to Dr. Frenk):

Just one remark with regard to the pitfall due to changes in glomerular filtration rate. I would like to state that recent studies have revealed that the excretion of enzymes and iso-enzymes in urine does not necessarily reflect the glomerular filtration rate. I would just refer to the proceedings of a conference on enzymes in urine, held in Rheinfelden (Dubach, 1968), and I think it might be quite informative for those interested in this question to consult this book.

R. G. Whitehead:

May I say something about biopsy? The clinicians with whom I am associated would never agree easily to biopsies, certainly not serial biopsies in the field.

This is a problem because obviously biochemists want to work on tissues. Biochemistry is a science of cells, not of urine! Our tissue studies are confined mainly to experimental animals. From these we attempt to establish relationships between cellular changes and biochemical abnormalities in serum and urine, etc. In this way, we hope to be able to show that the latter have relevance in terms of cellular metabolism.

Glomerular filtration rate: May I just say something about that? The importance of glomerular filtration is one of the reasons why I favored expressing hydroxyproline excretion in relation to creatinine. My assumption was that, if glomerular filtration rate was reduced, it would reduce the excretion of hydroxyproline to the same extent as it would reduce the excretion of creatinine.

1.5. Relative Distribution of Amino Acids

A. von Muralt:

It might be worthwhile to say a few words about the relative distribution of amino acids.

It is interesting to see that the cystine/glycine and valine/glycine ratios seem to be rather good indices of PCM.

A. E. Harper:

There are substantial differences in the distribution of free amino acids in different organs and tissues, and great differences in the sensitivity of organs to changes of diet.

Changes in plasma and muscle tend to be similar but the concentrations are very different, muscle concentrations of individual amino acids being from two to ten times those of plasma. The pattern in the brain is distinctly different from that of most other tissues, and again is much less sensitive to changes in circulating amino-acid concentrations, owing possibly to the blood brain barrier. The liver seems to be less sensitive to changes in the dietary pattern than plasma or muscle.

Examination of the total pool (concentration times volume) gives a still different picture. The plasma amino acids represent a minute fraction compared with the amounts in other tissues. Plasma should, therefore, provide a relatively sensitive index of changes occurring in other tissues, because a small change in a large tissue would be reflected in a relatively large change in plasma. I think this is what we see in our amino-acid imbalance studies.

We need more investigation of the links between changes in the metabolic activities of individual tissues and their reflection in the plasma amino-acid pattern to enable us to interpret these changes properly.

G. Fanconi:

I was very interested to see that the level of some amino acids is increased if the food is imbalanced in amino acids. This gives perhaps an explanation of the observation of Dubois et al. (1959). When kwashiorkor children were fed a diet with sufficient proteins, during the first days and weeks the aminoaciduria was very high. Probably because some of the amino acids which had a high level in the body were eliminated in the urine.

A. E. Harper:

Our animals are well when we start the experiments and the experiments are short. The amount of free amino acids in the urine of rats fed a diet with an imbalance increased above the control value, but even a 2 to 3-fold increase does not represent much loss. The renal thresholds for some amino acids can be exceeded by feeding a large quantity of a single amino acid but not with the quantities we use to create imbalances.

J. S. Dinning:

Did you make any calculations of the various amino-acid ratios ?

A. E. Harper:

The main calculation we make is the reverse of that made by Dr. Whitehead, the indispensable to dispensable ratio, because in our studies the indispensable amino-acid concentrations are elevated. We

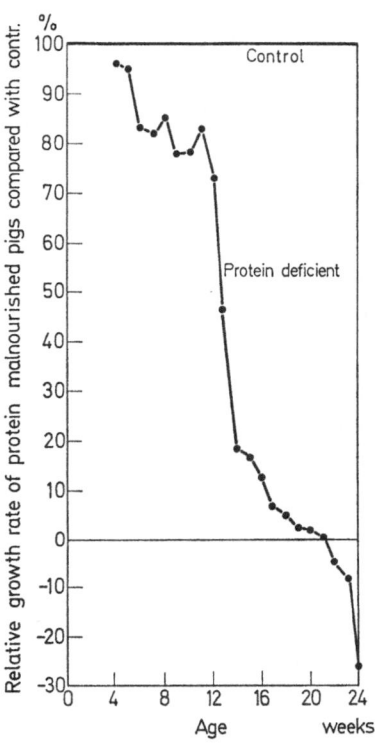

Fig. 3. Growth rate of pigs

have attempted to demonstrate a correlation between the ratios of indispensable to dispensable amino acids and changes in food intake. A high ratio tends to be associated with low food intake but not always.

J. S. Dinning:

This happened in few hours, whereas in children it is apparently necessary to develop the disease before it happens.

R. G. Whitehead:

I would like to show you some results of a pig experiment which shows the development of elevated amino-acid ratio in malnourished pigs, and also gives some information on the relationship between this metabolic abnormality and growth and development (Fig. 3). The amino-acid ratio is determined with a semi-quantitative paper chromatographic method. The numerator is glycine, serine, glutamine and taurine; the denominator is leucine, isoleucine, valine.

What we wanted to find out was: when does the abnormal amino-acid ratio develop? Has it any significance in terms of the nutritional status

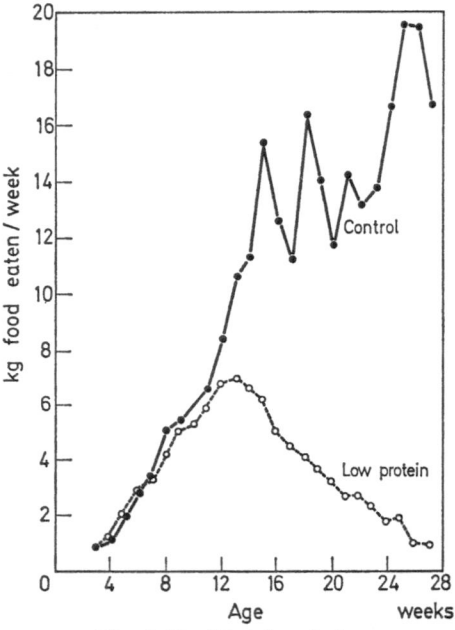

Fig. 4. Food intake of pigs

of the animal? The animals were fed diets containing gradually decreasing amounts of protein and increasing amounts of carbohydrate in the form of sucrose.

The animals initially grew fairly well in spite of the fact that the protein concentration of the diet was falling. But at a certain crucial stage (12 weeks), there was a very marked drop in the rate of growth. Eventually the animals started to lose weight.

Fig. 4 shows the total food intake of the pigs compared with the controls. Initially, in spite of the poor nature of the diet, the animals continued to eat the same amount as the controls, but at week 12 the animals started to lose their appetite. A loss of appetite is an important behavioral characteristic.

There was no significant change in amino-acid ratio at first (Fig. 5), but at 12 weeks the amino-acid ratio rose significantly. I think these experiments do show that an elevation of the amino-acid ratio does reflect profound changes in the well-being of the animal. The distortion of the serum amino acids occurred at the same crucial time as the development of the anthropometric and behavioral abnormalities (Figs. 3 and 4).

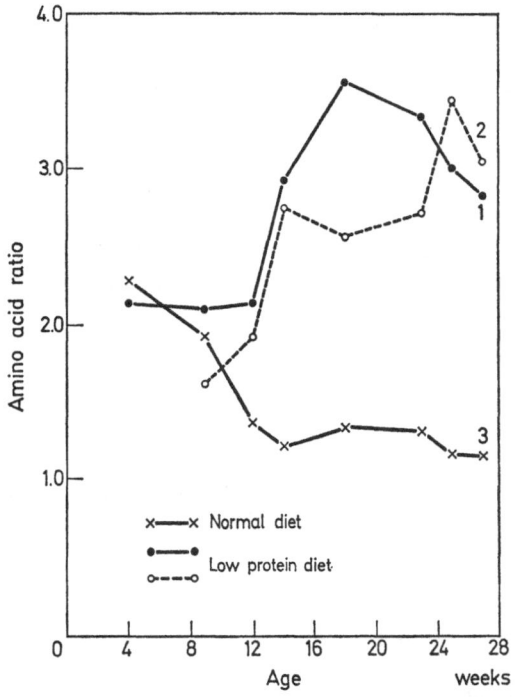

Fig. 5. Changes in amino-acid ratio in protein malnourished pigs

J. S. Dinning:

What percentage protein were they getting in 10 weeks?

R. G. Whitehead:

It was about 4% protein. The ratio only rose when nutritional stress was really very severe.

C. Gopalan:

There is one human situation in which we feel that amino-acid imbalance may play an important part. In Hyderabad, we are confronted with the problem of *pellagra*. Nearly 1% of our hospital admissions are cases of pellagra. The pellagra in our situation is not a disease of the maize eater, it is a disease of the millet eaters. Millet is not deficient in tryptophan, and the nicotinic acid is in an available form.

Animal experiments showed that excess of leucine present in millet may be responsible. In cases of pellagra, the administration of leucine can aggravate the clinical manifestations, especially the mental manifestations. We can produce the most striking electroencephalographic abnormalities in pellagra patients by giving them leucine only. We can produce marked distortion of nicotinic acid metabolism, as observed by the excretion of nicotinic acid products in the urine, by feeding leucine to human volunteers — or to rats, such as increase in the quinolinic acid and a marked decrease in the 6-pyridone in the urine. In pups, by adding leucine to a diet which is otherwise non-pellagragenic, you can induce blacktongue, blacktongue in dogs by feeding them a millet diet.

We believe that, in the millet eaters of the Deccan Plateau in Central India, we are perhaps dealing with a situation of amino-acid imbalance. If this were true, I think this would be the first instance in which an amino-acid imbalance as such is being shown to be responsible for a real important public health problem of that magnitude.

1.6. Skin Studies

R. G. Whitehead:

It is very surprising that, with abnormalities in the skin being such an important clinical feature of kwashiorkor, it received so very little biochemical attention.

S. Frenk:

Our own studies on skin were focused on composition of the subcutaneous layer, and actually are not relevant to this point (Frenk, Metcoff, Gomez et al., 1957).

Very little seems to be known about the origin of *hyperpigmentation* of the skin in malnourished subjects. One possible approach to this question may be via serotonin metabolism. Since serotonin metabolites appear to be low in blood and in urine (Valenzuela, Hernández-Peniche and Fernández-Varela, 1966), a defect in melatonin synthesis may be postulated as a hypothetical explanation for skin hyperpigmentation in malnutrition.

R. G. Whitehead:

I would like to say that Dorothy Coward, one of my colleagues, has been working on the metabolism of skin collagen in malnourished rats.

The amount of collagen beneath a standard area of skin is lower than in normal rats of the same age. We have also fractionated the skin collagen into the various soluble fractions and we found that the concentration of 0.15 M NaCl collagen, i.e. the most recently formed metabolic component of collagen, is much reduced in both protein deficiency and undernutrition.

There is a high statistical relationship between the concentration of these soluble components and the amount of hydroxyproline excreted in the urine. Thus, in rats at least, the low excretion of hydroxyproline reflects a reduced rate of synthesis of collagen.

The reason that there is less collagen in the skin of the malnourished animals is not because the actual collagen concentration falls, but that it does not increase at the same rate as it does in a normal rat. In a normal rat, the concentration of collagen increases as the rat gets older but in a protein-malnourished rat, due to the reduced rate of synthesis, the amount of collagen is relatively less; so, at a given age, there is less collagen in the skin than one would expect.

2. Anthropometrical Data

The *weight for age* is undoubtedly the best field test in community surveys at the present moment, approximate as it may be. The weight for age, though, is very difficult to evaluate in areas of the world where the age of the subjects is not accurately known. Biochemical tests have a definite advantage in that they are relatively age-independent.

D. B. Jelliffe:

We include six measurements in field surveys: weight, height, the arm circumference, the triceps skin fold, the head circumference, and the chest circumference.

These measurements do not cover all body dimensions; laterality, for instance, is not being measured at all. However, we have to try to limit the number of simple tests to those which give us as much information as possible within a limited time.

A. von Muralt:

One could refine the morphological measurements by measuring cell size and cell number with the microscope in tissue-samples. Actually the weight is only the integrated sum of what goes on in the body with respect to cell size and cell number.

S. Frenk:

The effect of malnutrition on tissue size may be due either to a reduction in cell number or to a decrease in cell size. If kwashiorkor was characterised mainly by the latter phenomenon, this might explain the rapid catch-up growth of the affected children, once they are free from infection and successfully re-fed.

G. Arroyave:

Recent studies in rats show that the rat completes its number of cells in the brain around the age of 16 days. Careful studies have shown that any damage or insult before the 16th day gives *permanent damage*, but after 16 days it only gives damage from which the rat can recover.

This age in the human corresponds to about 5 months. Therefore, if this is true, it would be tremendously significant whether the insult happens at an early age or at the age of 1 to 2 years, especially with regard to mental development. The studies were done on the number of cells in the brain and in other organs.

C. Clinical and Socio-Economic Aspects of PCM

1. Kwashiorkor and Marasmus

1.1. Presentation of the Data on the Prevalence of PCM

J. M. Bengoa:

There is a need for better information on the prevalence of malnutrition in the world because until now the data available are really very scanty and not very well collected and analysed. We should improve, in the future, our presentation of the data on the *prevalence* of protein-calorie malnutrition. Marasmus is a very long disease which may take 8, 10 months or even a year, while kwashiorkor is an acute condition in which the child dies or survives; in 2 months the clinical situation is decided in one direction or in another.

In this case, let us assume that in a field survey for a given day there are four cases of marasmus and two cases of kwashiorkor. Some people suggest that the number of cases of malnutritional marasmus is double that of kwashiorkor, and this is true in the above example from the point-prevalence point of view. But if we assume that the course of a kwashiorkor is 2 months and we have two cases in January, two cases in March, two cases in May, etc., as a whole, we have 12 cases within the year, against 4 cases of marasmus. Therefore the yearly prevalence is 12 cases of kwashiorkor versus only 4 cases of marasmus (Table 2).

Incidence is quite different! The incidence of kwashiorkor and marasmus in this example in January was four cases of marasmus and two of kwashiorkor; but in February, it was zero for marasmus and zero for kwashiorkor. Incidence reflects only the new cases in a given period of time.

You can see that the three interpretations of the data — point prevalence, yearly prevalence, and monthly incidence — are completely different! Therefore it is very difficult to say in a given period if there are more cases of marasmus or more cases of kwashiorkor.

Table 2. *Theoretical example to illustrate differences between prevalence and incidence in nutritional marasmus and kwashiorkor, and interpretation of data*

Disease	Months												Yearly prevalence
	J	F	M	A	M	J	J	A	S	O	N	D	
Nutritional marasmus	4												4
Kwashiorkor	2		2		2		2		2		2		12

Interpretation of the data

Point prevalence: 4 cases of marasmus versus 2 of kwashiorkor.

Yearly prevalence: 4 cases of marasmus versus 12 of kwashiorkor.

Monthly incidence: in January, 4 cases marasmus versus 2 kwashiorkor,
in February, 0 cases marasmus versus 0 kwashiorkor,
in March, 0 cases marasmus versus 2 kwashiorkor.

C. Gopalan:

If kwashiorkor and marasmus were two entirely different syndromes, it would be necessary to know the relative incidence. But the point which I wish to emphasise here is that, if both syndromes are due to the same dietary background, it is not so essential for us to get into an argument as to whether kwashiorkor is more frequent than marasmus or marasmus than kwashiorkor, because both of them are due to the same cause. I feel we have sufficient evidence that marasmus as seen in the preschool child and kwashiorkor are both manifestations of one problem and are related to the same dietary background of protein-calorie deficiency.

D. B. Jelliffe:

One of the things which has come out of this meeting is that perhaps we should be wise not so much to divide into kwashiorkor and marasmus as into *infantile* and *post-infantile disease*. The etiologies so frequently are different in these two age groups. The infantile age group is particularly related to lactation failure.

Also, when undertaking a community point-prevalence survey, one should always try to get complementary data which will give at least some idea of what has been happening during the whole year in that particular community. The usual way one can do this is by looking into the data from hospitals, health centers and the like of the particular region, realising of course that these are biased statistics. Prevalence must be complemented by an approximate idea of incidence.

1.2. Differences between Kwashiorkor and Marasmus

G. Arroyave:

We find in our marasmus, and I stress *our* marasmus, that the total concentration of amino acid is low, almost as low as in our kwashiorkor. The pattern of changes in marasmus is very similar but not so intense as in kwashiorkor.

I think that we are over-simplifying the problem when we refer to marasmus and kwashiorkor so specifically. Animal experiments suffer from the fact of extreme over-simplification of the human condition. In humans, we do not find the same pure protein or pure calorie deficiency that we provoke in animals in the laboratory. We always find a combination of both conditions. Our marasmic children are suffering from a relatively low intake of calories and a very low intake of proteins. The end result with regard to the body metabolism is a protein deficiency.

If you have a primary calorie deficiency, the proteins are used as calorie sources. The body suffers from a shortage of amino acids for protein synthesis, a shortage which is of secondary origin. We should talk of a *severe syndrome of protein-calorie deficiency*, without trying to differentiate between marasmus and kwashiorkor.

The only person who can tell us what is happening to any specific child is the biochemist. This is why I insist that not one single method can give us the answer, but that we have to apply a "biochemical battery" of determinations which are complementary to one another.

S. Frenk:

On the High-Plateau of Mexico, clinical features of protein-calorie malnutrition in children of more than 1 year of age differ from those seen in tropical zones, mostly by a lower incidence of hepatomegaly, even if liver steatosis is of a similar degree.

As pointed out before, when talking about marasmus one could perhaps make a distinction:

At one end of the scale, cases of fetal malnutrition, born with all the clinical and at least part of the biochemical features of the breast-starved or otherwise starved small infant.

At the other end, the marasmic pre-school age children, whose *universal signs* in the sense of Ramos-Galván (1958) are the same as those of the fat-looking oedematous patients. With regard to water content and other physico-chemical features in this group, the two clinical types are indistinguishable.

One may say that, except for pure starvation, incipient malnutrition may easily be recognised in the field by the appearance of cheeks which look too big for the chronological age of the subjects or their body weight.

D. B. Jelliffe:

As many of us will know, the problem of the so-called "stuck marasmus" is very important in tropical pediatrics. We worked out in Kampala that it cost something of the order of 300 dollars to treat a child with marasmus, partly because it took such a long time.

I feel, however, that this illustrates again the difficulty of generalising from one part of the world to another. Kwashiorkor varies as a syndrome, but so does marasmus. The younger the marasmic patient is, the more difficult is the treatment, and the more uncertain the long-term prognosis. This may perhaps be related to the fact that the critical period for permanent damage to growth potential is in the earlier months of life.

One of the problems that exist in treating marasmus in a young child, as compared with treating a similar situation in an adult, is the *size of the stomach*. We are faced with the sheer physical problem of how to get sufficient protein-calories into the relatively small stomach of the young child. This is where intra-gastric drip feeding plays a useful part and where the use of compact calories in the form of oil and sugar can be very beneficial.

A. E. Harper:

Are there obvious differences in the *appetites* of patients with the kwashiorkor-type and the marasmic-type?

R. G. Whitehead:

It is difficult to answer this question in children, but we can answer it from studies on experimental animals. Animals which have been deprived of food, and are dwarfed, eat as soon as food becomes available and grow very rapidly. The ones which have been severely protein-malnourished take a little more time to regain their appetites.

D. B. Jelliffe:

There is a considerable difference in the appetite between the two in my experience.

The kwashiorkor case, as we all know, is withdrawn, miserable and apathetic, whereas the marasmic child shows much more interest in the outside world and much greater appetite. But it is difficult to compare the two, because they usually cover two different age ranges.

G. Arroyave:

The experience of INCAP, very consistently confirmed, is: that a child with kwashiorkor is apathetic, has more psychological disturbances and has a much poorer appetite than the marasmic child. The marasmic child is anxious to eat and from the very beginning starts eating well.

One of the problems in slowing the recuperation of the marasmic child is the intake of calories. If the requirement of the child by his age is 100 calories per kilo, you have to raise these to 200 and 250 if possible; then his weight goes up very fast.

H. Isliker:

I noticed the extreme sensitivity of the test measuring the *ratio of cysteine to glycine* in kwashiorkor. I want to ask you whether such a test has been carried out in marasmus as well, where the amino-acid pattern apparently is normal.

G. Arroyave:

Table 3 includes two cases of *marasmus* and the ratio is low compared with normal children, but it is even lower in kwashiorkor. In marasmus, there is an altered plasma amino-acid pattern, almost as much as in kwashiorkor.

C. Gopalan:

We had cases of kwashiorkor in which, after the first 2 or 3 days of rehabilitation, the appetite was restored. We had other cases in which the appetite was not restored after several days of treatment and it was in this latter group that the fatalities were high. We had cases of marasmus where the appetite was good and cases in which the appetite was poor. I may add here again that we use the word "marasmus" in a very much wider sense to cover the children of the older age groups who were emaciated but who did not have oedema.

Dr. Whitehead, you referred to cases of kwashiorkor in the adult. I would plead for the restriction of the term "kwashiorkor" specifically to the pre-school children.

Table 3. *Plasma cystine and glycine in population groups having different nutritional characteristics*

Group	No.	Cystine (mg/100 ml)	Glycine (mg/100 ml)	Ratio
1 Pregnant women Guatemala City (UIU)[a]	5	0.576	1.210	0.516
2 Pregnant women S. Ant. P. (LIR)[a]	6	0.295	1.599	0.208
3 Non-pregnant women S. Ant. P. (LIR)	7	0.763	2.643	0.289
4 Newborn children Guatemala City (UIU)[b]	5	1.141	2.531	0.472
5 Newborn children S. Ant. P. (LIR)[c]	6	0.865	2.970	0.290
6 Well-nourished children 3 to 6 years of age	5	0.736	1.607	0.471
7 Children with kwashiorkor 2 to 6 years of age	6	0.113	1.577	0.069
8 Children with marasmus 1 year of age	1	0.163	1.266	0.129
	1	0.269	1.596	0.168

[a] UIU — Upper income urban; LIR — Low income rural. Ninth month of pregnancy.

[b] Cord blood from group 1.

[c] Cord blood from group 2.

We should not apply the term "kwashiorkor" to the adult cases who constitute an entirely different category.

1.3. Recovery from and Prevention of Kwashiorkor and Marasmus

J. M. Bengoa:

From the biochemical point of view, marasmus in small children is the same as starvation in the adult. If it is true that it is the same syndrome, why then in children is the recuperation from marasmus so difficult, in contrast with the quick recuperation from starvation in the adult?

R. G. Whitehead:

This is a very interesting question. In Uganda, marasmic children tend to recover much more slowly than those with kwashiorkor.

D. B. Jelliffe:

I think it means that we must re-emphasise even more the necessity of introducing adequate quantities of available foods to young children in the *second 6 months* of life to ensure adequate intakes of both protein and calories.

We realise more and more that the PCM situation in a community is very much related to the staple of the community but not only, as the classical thinking was, because of its protein composition, but also because of its calorie content. In other words, a community which, as in parts of Uganda, subsists on a staple which is low in protein and has a calorie content diluted by bulk, water and cellulose, is at a double disadvantage: low protein and a diluted calorie content!

C. G. King:

The focal point in the total protein problem seems to me the normal transition from weaning time to a mixed diet. It should be strongly emphasised in all education programs that shortening of the period of *breast-feeding* is one of the great tragedies. We must work vigorously, particularly among pediatricians and mothers of children who are faced with the problem, to maintain a reasonable degree of breast-feeding, as being essential to the best human development.

Our wide growth of baby foods on the whole is good and can be constructive but, if it results in shortened and sometimes completely omitted breast-feeding, it is very unwise.

The development of the central nervous system is crucial within the first year of life. This has been neglected as a field of intensive research too long and we need carefully controlled work with experimental animals as a guide.

The tendency to develop baby foods as well as adult foods on the basis of flavor is one of our hazards. The flavor of foods is gaged largely by tasting panels of adults, preferring foods with more salt than the human body needs. A present trend among baby-food manufacturers is to include from 5 to 10 times more salt than the salt content of breast milk from a well-nourished mother.

E. M. Mrak:

It has been my understanding that at least some of the companies, in setting up the panels, are trying to cater for the tastes of the mothers rather than for the babies, so probably what we need to do is to educate the mothers to accept something different! The second thing I was wondering about is breast-feeding. If the native mother does not give up breast-feeding early, she loses her husband to someone else!

Producing a cheap ready-made food successfully depends on the economic situation in the country. If we are not going to produce something that people can buy, there is no help!

D. B. Jelliffe:

The commonness of protein-calorie malnutrition needs emphasis. For example, in a survey in Haiti, 7% of the pre-school children had kwashiorkor, 40% of the children by anthropometric measurements were below the third percentile and 60% of the children in the community, taking random samples in the whole country, were mildly, moderately or severely malnourished.

I would like to make one comment on Dr. Bengoa's paper. He made many interesting points and one of these is the burning question of *decelerating the process of early weaning*. I would like to draw people's attention to the strange situation which has arisen in many parts of the world in which "mammal" man is ceasing to breast-feed his children. The change from breast-feeding to artificial feeding has become a world menace invading many areas in less developed parts of the world — a nutritional retreat rather than an advance.

There are certain problems here. Firstly, how do we get over the realisation that successful artificial feeding in developing regions is often impossible for economic, hygienic and educational reasons? Marasmus and infectious diarrhoea are extremely likely to occur.

Secondly, how do we persuade public health and nutrition planners that breast milk indeed is a food? This is strangely difficult, as it is neither bought nor grown, nor even seen if it is successfully used. It is strange that there is a tremendous amount of work concerned with producing new protein foods, and yet mother's milk — this time-tested, consumer-approved protein food — is tending to disappear.

A third question: how do we persuade people of the need for research into lactation failure in developing countries and into the methodology of remotivating communities towards a breast-feeding?

Lastly, do people really appreciate the economic and food production consequences of this trend? It has been calculated that if in India the present pattern of lactation were to cease, and artificial feeding was to become widespread, this would require the milk production of at least five million lactating cattle to make this loss good.

I do feel that the point Dr. Bengoa has made here, of decelerating the process of early weaning, is one of immense importance! It is not the only problem in dealing with the question of protein-calorie malnutrition, but it is one which I think is under-emphasised.

2. Instinctive Choice of the Right Food

D. Bovet:

Another aspect of the homeostatic equilibrium was demonstrated in biology in the classical experiment of Richter (1942), who showed that the rat in presence of a choice of diets — it was the so-called "cafeteria technique" — is able to obtain a best equilibrated diet and even, in some pathological states, as for example parathyroidectomy or adrenalectomy, the rat is able to change its diet, to adjust it to a normal equilibrium.

This was made 20 years ago and I think it was rather successful because it was proved that even children are able to make the choice, just as the rat. It would be interesting to make some experiments in this direction and to try to give very young children the possibility of a choice between various types of food.

A. E. Harper:

Work on this subject suggests that, if the physiological response to the dietary component is rapid, the animal makes a choice quite effectively as for protein, amino acids or thiamine. The ability to select for components for which the physiological response is much slower, such as vitamin A, is not very great.

3. Central Nervous Response to Nutrition

D. Bovet:

It is known that the brain tissue has a very high protein turnover, and Dr. von Muralt spoke about the shift between the protein metabolism in liver and in muscle, but we know very little about the shift between the protein metabolism in other organs. The brain is certainly very sensitive to a change in the diet, especially during growth, in the first part of life.

The protein turnover in brain is high, as was demonstrated by Gaitonde (1953, Beloff-Chain (1955), Richter 1962), and by many others recently. Now we have a very limited number of experiments in this field, but a Czechoslovakian colleague, Lát (1960, 1965), was able to observe that rats which receive plenty of milk were more responsible to stimuli. He also demonstrated that rats with high-carbohydrate diet were more exploratory and active than rats with low-carbohydrate diet, and also that rats which are more active also introduce with the "cafeteria technique" more carbohydrate in their diet.

We have also a stimulating observation of Cowley and Griesel (1966), who observed that giving an inadequate diet to an albino rat, where the main deficiency was a protein content of only 21%, provoked a diminution in the learning capacity of the rat within a month.

Now, as Dr. Fanconi knows, it is an old joke in my country that the people of Florence say they are more intelligent because they eat a special kind of large white bean, and that is the reason for the extra-ordinary development of Florence during the Renaissance.

A. E. Harper:

There are some observations on the *running ability of rats* fed on two different levels of protein in the diet. The test consisted of an exercise, programmed so that the work gradually increased with time. The rats fed on the low-protein diet had no problem in completing the entire program plus doing sudden spurts of exercise. Those which had the high-protein diet were not only unable to do the spurts, they were actually unable to complete the full exercise procedure that had been outlined.

The rats on a low-protein diet were not severely depleted, but they were lean. Those which had the high-protein diet tended to be somewhat obese and somewhat larger.

Dr. Fanconi suggested that aggression is a result of full nutrition. I would suggest that aggression is primarily the response to an inability to satisfy basic biologic needs or efforts to suppress the demand that such needs be satisfied because the demand has appeared in a way that has created fear in those who can satisfy the need. This is seen very clearly in the inability to satisfy the need for food. Lepkovsky (1953, 1959) has described this vividly as the development of an ugly personality and it can bring forth an ugly response. Cooperative human relationships also seem to be a basic biologic need which must be satisfied by eliminating social, economic, religious and other prejudices if we are to solve the world's food problems without resorting to aggression.

C. Gopalan:

I would like to endorse the plea that what is needed is perhaps a compassionate and not too myopic view of what is happening in the so-called developing countries. I am now referring to statements like "the intellectuals of developing countries do not cooperate". I think this statement is not necessarily true and in any case the attitude of non-cooperation is not necessarily the special attribute of the intellectuals of developing countries alone. The youths of the whole world are in revolt and you see exactly the same sort of lack of cooperation in the streets of Paris as you see in the streets of Hyderabad.

We should not confuse civilisation with technical development. We should not talk of them as though they were equivalent and synony-mous. There are many countries which are technically backward, but which are certainly not uncivilised. It is very essential that the leaders in the so-called developing countries understand these factors. Failure to understand this is often the real barrier to the eradication of malnutrition in large parts of the world.

4. Socio-Economic Aspects

G. Arroyave:

Since we are talking about the causes of malnutrition, we are getting into very complex grounds: food processing, food industry, education and other factors. One of them is the socio-cultural characteristics of the

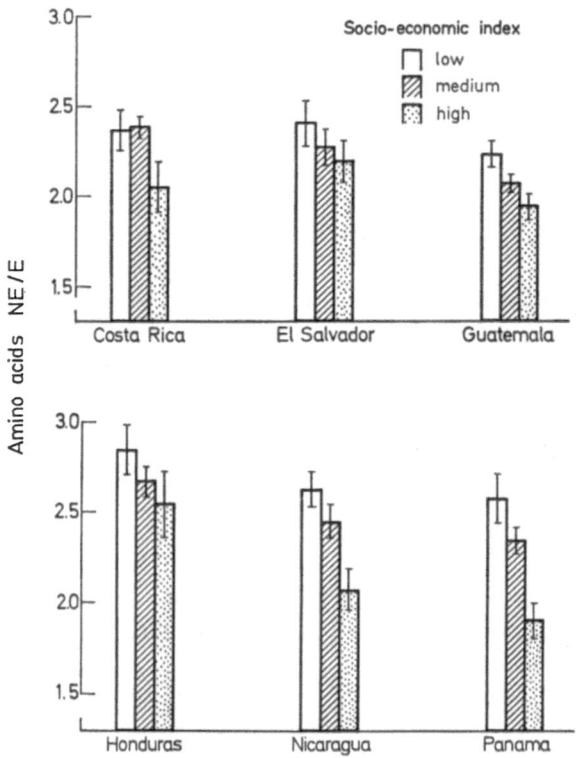

Fig. 6. Nutritional survey in Central America and Panama. Serum amino-acid ratio NE/E in the rural area of Central America in relation to the socio-economic index of the families. ♂

population in which malnutrition exists. It has been said that malnutrition is a *social disease*. The relationship between the socio-cultural level of the family or of the community with the prevalence of malnutrition has been recognised, but mostly on an empirical basis.

We collected information in a recent extensive nutrition survey in Central America on the socio-cultural level of the families, using a quantitative index based on eight criteria of socio-cultural development. We could then scale the families with indices from one to three, one being the lowest and three the highest socio-cultural index. This included housing, number of rooms, number of persons per room, number of beds per person, water supply, disposal of wastage and feces, property of

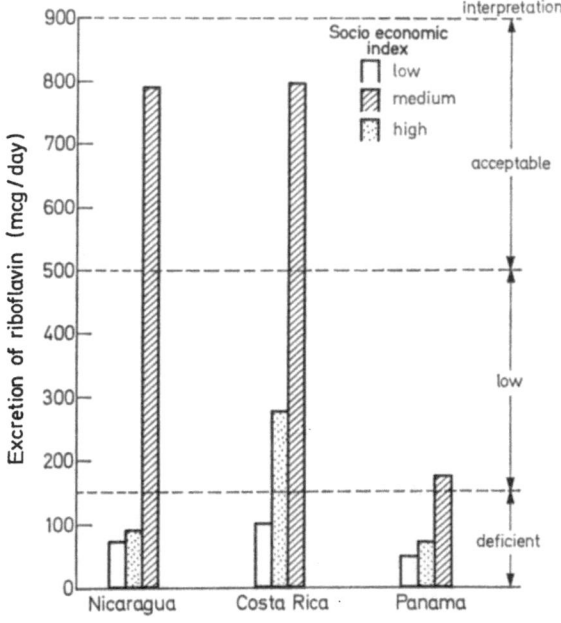

Fig. 7. Nutritional survey in Central America and Panama. Urinary excretion of riboflavin in the rural area of Central America in relation to the socio-economic index of the families (1965—1967). (♂, 0—4 years of age)

animals and land, etc. Using this index, we could then classify some 3000 families.

Using this index, we classify the families in the low, the medium and the high socio-economic index. In Fig. 6, the non-essential to essential amino-acid ratio of *Whitehead* is represented.

It shows very clearly that, for each one of the six Central American countries, the ratio is lowest or closest to normal in the upper socio-cultural group, and abnormally high in most of the low socio-cultural family groups.

The riboflavin excretion in the urine (Fig. 7) is interesting because this vitamin is contained in the same foods which contain animal protein or good quality protein. It is useful as an index of better nutrition. Take for example Nicaragua: this country is characterised by the highest supply of milk per capita and the highest average consumption of milk per capita. But look what happened with this supply of riboflavin:

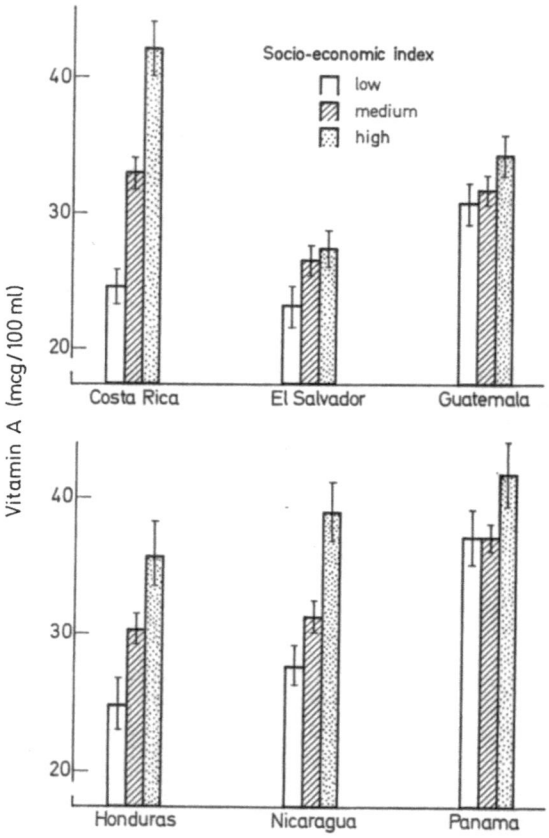

Fig. 8. Nutritional survey in Central America and Panama. Vitamin A in blood serum in the rural area of Central America in relation to the socio-economic index of the families. ♂

the low and the medium socio-cultural groups do not obtain much of this supply; it is only the upper socio-cultural group which consumes most of the milk. All of these are families in the rural area, cities were not taken into account. The same thing happens in Costa Rica; Panama is a country characteristically low in supply of riboflavin per capita and all groups in the rural area have a low excretion in the urine.

As to Vitamin A, we again can see a marked difference between the three socio-cultural groups (Fig. 8). Ascorbic acid is evenly distributed among the three socio-cultural groups (Fig. 9). In all instances, the level or the concentration in the plasma is adequate for any of the socio-cultural groups, even the first one. We ask ourselves: why?

Ascorbic acid is contained in foods in Central America which are not only abundant but of very low cost. In fact, many of the sources like

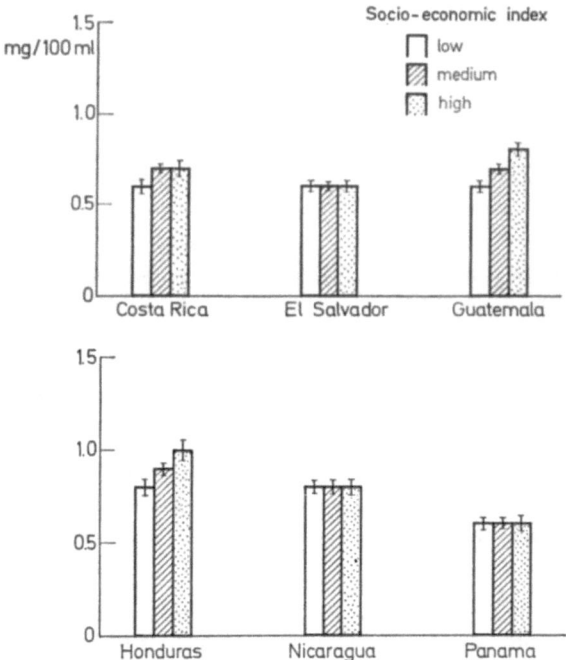

Fig. 9. Nutritional survey in Central America and Panama. Ascorbic acid in the blood plasma in the rural area of Central America in relation to the socio-economic index of the families (1965—1967). ♂

mangoes, tropical fruits, have practically no market value, so everybody can get his share of this supply of vitamin C.

On the other hand, riboflavin, protein and vitamin A are contained in foods which are expensive, even when they are abundant. The calculation shows that the supply of 60 g of protein per day for an individual in Central America would cost, if coming from corn: 28 cents; from milk: 40 cents per day. The supply of ascorbic acid, 40 to 50 mg per day, costs less than one cent and in some instances it is free. It is useless to produce very nutritious foods for a world that cannot buy them.

D. B. Jelliffe:

At the same time, there are parts of the world where people of relatively low income are already buying foods. For example in the Caribbean, the pattern of purchasing food from shops and stores has already become established. Under these circumstances, what one requires is food which is more appropriate to the local community's needs and purse, but which will have a reasonable nutritional value in relation to cost and yet be of high prestige.

C. G. King:

In a city area there will be an appreciable proportion of the popula ti n with an income high enough to buy good quality food. It is worth encouraging the entry of such food into the developing countries, even though the market is going to be slow to develop and, in the meantime, one should try to develop patterns of protein intake from the low-cost foods that are adequate.

In Haiti, they have practically eliminated kwashiorkor in sections where it was formerly at its worst, with simple combinations of the right kind of beans at the level of 30% by weight, with corn, sorghum or rice, as they prefer.

C. Gopalan:

In most Asian countries, where the great percentage of the populations lives in rural areas, recourse to processed foods on a very large scale will, if anything, increase the protein gap as far as the rural population is concerned.

Under such circumstances, emphasis should be laid on the utilisation of the foods which are produced in the villages, rather than on attempting to transport them to the towns where they can be processed and then shipped back, running into inevitable bottlenecks with regard to storage, transport, etc.

D. B. Jelliffe:

I agree very much with Dr. Gopalan. There sometimes tends to be a pseudo-dichotomy between the nutrition education people — "let us use what is locally produced to the best purpose" — and the food-processing — "let us make an infant food".

There can be place for both of these. In some places the emphasis may be completely on one or the other or, very often, a differing blend of the two. The weightage that should be given to each of these particular approaches has to be geared to local circumstances.

C. G. King:

We are going to have a revolution in the problem of feeding low-cost sources of good quality protein by new types of corn which are being developed.

The protein quality is about the equivalent of casein, with high yield, high quality corn. They have cured kwashiorkor with it, and there will be a further advantage in feed for swine and poultry.

D. Single Cell Proteins

J. S. Dinning:

This matter of single cell proteins is one of great interest.

I would like to ask Dr. Rey if any studies about acceptability of any of these products, which might be better than algae, are known. Maybe it was the color that bothered people?

L. Rey:

Dr. Dinning referred essentially to the work done on chlorella, some years ago, and which is still going on. This type of algae presents a different problem. They are less digestible than other single cell proteins, due to the cell membrane itself, and they have been found to be a little toxic — not very toxic — but there is some toxicity built in most of this algae material. From the organoleptic side, they are appalling, which is a great difficulty. We have not only to manufacture a material which is nutritionally valid and which can be manufactured at low cost, but it has to be also pleasant in taste and all other aspects. All efforts to get rid of this difficulty at a limited cost failed.

Spirulina, which grows naturally as a little filament at the surface of Lake Chad, can be eaten direct and is in fact eaten by the natives, and has been eaten for centuries; this is a rather rich material: it contains around 60 to 65% protein; this material has apparently, I should not say, no taste, but is almost bland. This can be a new approach to algae. The interesting point in this spirulina is that it grows on carbon-dioxyde media, but unfortunately it has a slow growth, which is not economic.

G. Arroyave:

Dr. Scrimshaw, from the Massachusetts Institute of Technology, has obtained some results (still unpublished) of his testing of cellular proteins in young adults, and he emphasises the difficulty due to the high content of nucleic acids in the proteins that are produced.

Apparently, they result in a very high concentration of uric acid in the blood, with some dermal reactions on the hands and feet.

L. Rey:

This shows again that it is extremely difficult to forecast future applications to human beings by animal studies. The animals have a uricase which can very easily break down the uric acid, but the primates have none.

The allergic reactions are a problem of individual susceptibility; in some cases — as Dr. Scrimshaw mentioned — this reaction on the skin was extremely severe, with even a kind of bleeding of the surface, which disappeared in a couple of days, despite the fact that the diet was maintained. It is difficult to know whether these signs are due to the increase of uric acid or to some other compound, producing a specific allergic reaction.

D. Bovet:

What will be the interest in yeast and bacteria with regard to their vitamin content? Is it economically important to have the vitamins in these cells?

L. Rey:

The content in vitamins is extremely high in both compounds, but generally in favor of the bacteria. But it depends upon the different strains. With respect to vitamins, there is a certain plus in single cell proteins.

On the industrial level, it is much more difficult to grow bacteria in bulk than to grow yeast, for reasons of purity. The bacteria grows generally around pH7, the yeast grows at a very acid pH, around 4, so that contamination is much less common in yeast fermentation. With bacteria, there is in addition the problem of bacteriophage, which means that you are faced with the problem of having strains which are resistant to bacteriophage in order to start again. This is the main difference between yeast and bacteria. But on the vitamin side, the advantage would be for bacterias.

It is quite obvious that the Nestlé Company is looking forward to marketing this product one day, at least as a raw material. That does not mean that in their joint venture with Esso they will keep for themselves the right to use the basic material. They have been deeply concerned by the acceptability of these products and the ways to use them.

It is difficult to change the eating habits in most of the countries of the world, and to introduce something which is completely different in flavor and in structure would be almost impossible.

So the material should be bland. It should be mixed easily with conventional foods, in order to be at the disposal of the whole family. It should be incorporated in soft drinks, in soups or soup-spices.

C. G. King:

Has there been time enough to make exhaustive and really satisfactory testing on the risks of carcinogenesis or hormonal disturbances, in view of the possible abnormality of the hydrocarbons and derived products?

L. Rey:

We have already an experience of more than 3 years. Serious experiments have been done in an effort to check the induction of cancer and the teratogenic changes, as well as hormonal disturbances in animals, and up to now we have found no adverse effects.

E. Concluding Remarks

C. G. King:

In summary, there is and has been all along a general acceptance of the concept that protein-calorie malnutrition is the most crucial part of the world food problem. World-wide family planning in a vigorous and prompt manner is also essential for success in facing this as all other problems in man's struggle for a world-wide adequate food supply.

Specific issues that need emphasis in major national and international programs:

The importance of identifying actual food intake by specific individuals in each family and analyses of the food for specific nutrient value as consumed, followed by an appraisal of food available to the community.

Clinical examination by experienced and well-trained clinical personnel. It has the weakness of being subjective and difficult to express in quantitative figures for comparative data in different reports. The degrees of deficiency must be stated in general terms, whether there is primarily a caloric or an amino-acid deficiency. In most cases, it can be related to height and age, height and weight, skin conditions, oedema, and should be followed when possible by morphological studies.

Biochemical analysis of the blood plasma, urine and, if within the range for the study, specific biopsy. Tissues such as the skin, hair, blood cells, membranes of the mouth and so on will always yield valuable information. New methods of analysis, if possible multiple analyses by standardised methods, should become relatively cheap if one considers their value in terms of the expenditure in either money or time. We are entering a revolutionary age of using biochemistry in this kind of work.

Specific tests which seem to have great value from the general discussion: *hydroxyproline*, the *ratio of non-essential to essential amino acids, urea/creatinine*.

Other observations with respect to the background and case histories of the individuals — such factors as infections, age, and the medical records — add immensely to the biochemical data.

Intensive research to increase our basic knowledge of the environmental genetic and neurological factors that are related to protein-calorie malnutrition. The normal pathways of protein metabolism and the enzymic changes probably represent the most sensitive and quickly responsive fundamental changes that correlate with malnutrition.

We are going to find it necessary to evaluate the degree of reversibility as timed to a specific period of tissue development, most important in the brain tissue. That is why I was very glad to hear the emphasis on the need for more neurological work in nutrition.

Food Composition Tables of the Important Foodplants Used in West Africa

Table 1. *Index of names: scientific, English, French, German, Spanish and African.*
The reference number refers to tables 2 to 7

Scientific name	English name	French name	German name	Spanish name	African name	Nr.
Adansonia digitata	baobab, monkey bread	baobab, pain de singe	Affenbrotbaum	baobab	Ouolof: goni (tree) gif (grains) Bambara: sira Baoulé: fromdo Mossi: toega	31, 57, 79
Agaricus	mushrooms	champignons	Pilze	setas		
Allium cepa	onion	oignon	Zwiebel	cebolla		42
Amaranthus hibridus	amaranth, Chinese spinach	amaranthe, épinard du Soudan	Amaranth	bledo	Ouolof: m'boum, keur Bambara: moron'dyé Haoussa: nga léya	58
lividus	amaranth, Chinese spinach	amaranthe, épinard du Soudan	Amaranth	bledo	Ouolof: m'boum boudigen	59 a
spinosus	amaranth, Chinese spinach	amaranthe, épinard du Soudan	Amaranth	bledo	Ouolof: m'boum bou gor	59 b
Anacardium occidentale	cashew, common nut	noix d'acajou, pomme de cajou	Elephantenlaus	marañón	Ouolof: n'dakassu Bambara: finzan	32, 80
Ananas comosus	pineapple	ananas	Ananas	ananás	—	81
Annona muricata	soursoup, guanabana	corossol épineux	Anonen, Sauersack	guanábana	—	82
Aphania senegalensis	soapberry-tree	cerise du Sénégal		cereza senegalesa	Ouolof: khever Sérère: mbutj Diola: boul, koul	83

Table 1. (continued)

Scientific name	English name	French name	German name	Spanish name	African name	Nr.
Arachis hypogea	peanut, groundnut, goober	arachide, cacahuète, pistache de terre	Erdnuß	cacahuete	Ouolof: guerté Bambara: tiga Djerma: demfi Baoulé: dorou kouassi Haoussa: goudja	24, 60
Balanites aegyptiaca	desert-date, Egyptian myrobalan, soap berry, betuoil seed, heglik oil, sump oil	datte du désert	Wüsten-Dattel	dátil del desierto	Ouolof: soump Bambara: séguené Haoussa: adoua Mossi: tiegaliga Peulh: mourotauki, tané	33, 84
Banhinia esculenta reticulata	gemsbok-bean, gemsbuck bean, camels foot	banhinia	Banhinia	banhinia		61
Borassus aethiopicum	African fan palm, palmyra, palm	rônier, palmier à sucre	Palmyra Palme, Delebpalme	palmera de azúcar	Ouolof: sibi-ron Bambara: sébé Agni: koubé Kabyé: gbéou	85
Boscia senegalensis	aisen	boscia	Boscia	boscia	Ouolof: n'diandam, diendoum Bambara: béré Haoussa: ansa, dilo Mossi: nabédéga	34
Brassica oleracea	cabbage	chou pommé	Kohl	colinabo		43
Brassica oleracea botrytis	cauliflower	chou-fleur	Blumenkohl	coliflor		44

10*

Table 1. (continued)

Scientific name	English name	French name	German name	Spanish name	African name	Nr.
Cajanus cajan	pigeon pea, Angola pea, Congo pea, dhal, red gram, yellow dahl	ambrevade, pois d'Angola	Taubenerbse, Redgram	guandú	Témou: adoua	25
Capsicum frutescens	red pepper, tabasco	piment enragé, poivre de Cayenne	Cayenne Pfeffer	ají, var		45
Capsicum annuum	sweet pepper	poivron, piment doux	Paprika	ají, var		46
Carica papaya	papaya, papaw, pawpaw	papaye	Papaya	papaya	Ouolof: papayo Djerma: dandimoufa Bambara: manguié Kabyé: soumolou	86
Cassia tora (C. obtusifolia)	senna sickle, foetid cassia	casse foetide	Fetid-Gemüse	casia	Ouolof: n'dour Bambara: zélou Haoussa: tafessa	62
Ceratotheca sesamoides	false-sesame, false benniseed, honum, karkashi	ceratotheca	Karkashi-Samen	ceratoteca		63
Chrysobalanus orbicularis	cocoplum	mafuli, voratch	Ikakopflaume (Chr. icaco)	hicaco	Ouolof: voratch Sérère: vanora Agni: hanfuru	35, 87
Citrullus vulgaris (C. lanatus)	watermelon, kaffir melon, stockmelon	pastèque	Wassermelone	sandia	Ouolof: hatar (plant) beref (grains) Bambara: tsara Haoussa: gouna	36

Table 1. (continued)

Scientific name	English name	French name	German name	Spanish name	African name	Nr.
Cocos nucifera	coconut, coconut palm	noix de coco, cocotier	Kokosnuß, Kokospalme	coco		37
Cola acuminata	colanut, kolanut	noix de cola	Kolanuß, Kolabaum	nuez de cola		38
Cola cordifolia	colanut	noix de cola	Kola	nuez de cola	Ouolof: tabba, n'taba Mandingue: tabo Diola: boubamb Baoulé: oualé	88
Colocasia esculenta	taro, dasheen, cocoyam, arum lilly	taro	Taro, Coco-Yam	ñampi		6, 18
Corchorus olitorius	jute, bush okra, jews-mallow	corète potagère, jute, mauve des juifs	indischer Flachs, Blattgemüse der Jutepflanze	corete	Ouolof: m'bali Sénoufo: sobo	64
Cordyla africana (C. pinata)	bushmango, mu tondo	cordyle	Wildmango	cordilo	Ouolof: dimb, dimbv Sérère: når, sek Bambara: dougoura, dougouta	89
Crataeva adansonii crateva (C. religiosa)	crateva	crateva		crateva	Ouolof: horel, njorel Mossi: kalégain tohiga	65
Cucumis melo	cantaloupe, muskmelon, kaffirmelon, senatfruit	melon, cantaloup	Zuckermelone	melón tuna	Ouolof: yombe	48

Table 1. (continued)

Scientific name	English name	French name	German name	Spanish name	African name	Nr.
Cucumis sativus	cucumber, gherkin	concombre, cornichon	Gurke	pepino	Ouolof: nadio Bambara: djé	47
Cucurbita pepo	squash, field pumpkin, vegetable marrow	courge, courgette, citrouille	Kürbis	calabaza		49, 66
Cyperus esculentus	tiger-nut, earth almond, Zulu nut, rushnut, chufa	amande de terre, souchet comestible	Cyperngras	chufa		23
Daucus carota	carrot	carotte	Karotte, Möhre	zanahoria		50
Detarium microcarpum	sweet dattock	dankh		detario dulce	Ouolof: dankh Sérère: rahn	90
Detarium senegalense	Senegal dattock	détar	Talgbaum	detario senegalés	Ouolof: detar, ditakh Sérère: n'doy Bambara: bodo Malinké: bodo	91
Dialium guineense	velvet tamarind, white tamarind	tamarinier blanc	Tamarinde (weiss)	tamarindo	Ouolof: solom Diola: kossito Bambara: kofino Malinké: kofino Baoulé: krékré	92
Digitaria exilis	acha, acca, findi, fonio, hungry rice	fonio	Findi-Getreide	fonio		11

Table 1. (continued)

Scientific name	English name	French name	German name	Spanish name	African name	Nr.
Dioscorea alata	yam, water yam, white yam	igname blanche, igname de Chine	Yams	ñame	bokwa, bola, murum, chumbia, song, onthalai, lyawa	5, 16
Diospyros mespiliformis	medlar persimmon, monkey guava, swamp ebony	kaki de brousse	Kaki-Feige, Persimone, Tuki	palo santo	Ouolof: alom, aloume; Sérère: niantchiqué; Bambara: sounsoun; Baoulé: bablé goualé	93
Dolichos lablab (D. biflorus)	lablab bean	lablab	Lablab Bohne	lablab	Agni: guangono ahrua	26
Ficus gnaphalocarpa	fig	figue	Feige	higo	Ouolof: gang; Bambara: tourou; Sérère: deoun; Mossi: kankanga	67, 94
Ficus iteophylla	fig	figue	Feige	higo	Ouolof: lodo, loro; Sérère: mbélègne; Dogon: tegedu; Haoussa: schiria	95
Ficus platyphylla	broadleaf fig	figue	Feige	higo	Ouolof: m'badat; Bambara: ouan bolo; Malinké: n'kobo	96
Gynandropsis pentaphylla (G. gynandra)	African spiderherb, cats whiskers, bastard mustard, lerotho	gynandropsis		ginandropsis		68

Table 1. (continued)

Scientific name	English name	French name	German name	Spanish name	African name	Nr.
Hibiscus esculentus	okra, ladies finger	gombo, bonnet grec	Gombo	quimgombó		51
Hibiscus sabadariffa	red sorrel, Indian sorrel, roselle, rose mallow, soursour	oseille de Guinée, roselle	Rosella	rosa de Jamaica	Ouolof: bissab Bambara: da Haoussa: yakoua	52
Ipomaea batatas (Convolvulus batatas)	sweet potato	patate douce	Batate	batatas	Ouolof: patas Bambara: koudouba Barina: dantin Ewé: anago	7, 17, 70
Kerstingiella geocarpa	ground bean, potato bean, geocarpa, harms seeds	lentille de terre		lenteja	Bambara: dougoufolo Mossi: diguem tenguéré Bariba: soui	27
Landolphia heudelotii	gumvine, Guinea fruit	landolphia		landolfia	Ouolof: toll Bambara: gohine	97
Landolphia senegalensis		landolphia		landolfia	Ouolof: mad, mada Bambara: sagua	98
Leptadenia lancifolia (L. hastata)	leptadenia	leptadenia		leptadenia	Ouolof: talal, tiakhat Sérère: sarafat Sonrhai: anou	71
Lycopersicum esculentum (Solanum lycopersicum)	tomato	tomate	Tomate	tomate		53

Table 1. (continued)

Scientific name	English name	French name	German name	Spanish name	African name	Nr.
Mangifera indica	common mango	mangue	Mango	mango	Ouolof: mangaro Bambara: mankourou	99
Manihot utilissima (M. esculenta)	cassava, bitter roots	manioc amer	Maniok	yuca	Ouolof: gniambi Bambara: banankou Djerma: rogo Agni: agbo bédé	4, 15, 72
Moringa pterygosperma (M. oleifera)	horseradishtree, ben-oil tree, drumstick, never die	ben ailé, neverdaye, brède morongy	Meerrettichbaum	ben alado	Ouolof: neverdaye nébédaye Haoussa: elmaka Djerma: woundi-boundou Mossi: argentiga	73
Musa sapientium	banana	banane	Obstbanane	banana común	Soussou: dougou-fui Côte d'Ivoire: konadou Bambara: namassa	100
Musa paradisiaca	plantain	banane plantain banane à cuire	Mehlbanane	plátano		8, 19
Oryza sativa	rice	riz	Reis	arroz		2, 12
Parinari excelsa	parinarium, rough-skimmed-plum, Guinea-plum	parinaire		parinaria	Mandigue: mampata Sérère: lo Dioula: goulih	101

Table 1. (continued)

Scientific name	English name	French name	German name	Spanish name	African name	Nr.
Parinari macrophylla	Cayor apple, parinarium, ginger-breadplum	pomme du Cayor		manzana de Cayor	Ouolof: néou Sérère: daf Bambara: danga Haoussa: gaosa Mandigue: tamba-coumba	39, 102
Parkia biglobosa	African locust bean, nittatree	arbre à farine néré, arbre à fauve, mimos pourpre	Heuschrecken-bohne	algarroba africana	Ouolof: houlle Sérère: séou Bambara: néré Mossi: douaga Baoulé: kpalé	40, 103
Pennisetum spp.	millet unclassified	mil non spécifié	Neger-Hirse viele Arten	panizo var.		1, 9
Persea gratissima (P. americana)	avocado, alligator pear	avocat	Avocatobirne	aguacate		104
Phaseolus spp.	beans, all species	haricots verts, toutes espèces	Bohnen, alle Arten	judía var.		28, 54
Psidium guajava	guava	goyave	Guayave	guayaba		105
Ricinodendron heudelotii	manketti nut, musodo, African woodoilnut	ezezan, huile de Sanga	Issagnila Nüsse	nuez de ben	Côte d'Ivoire: akwi, hacbiouagpi	41
Solanum tuberosum	potato	pomme de terre	Kartoffel	patata		20

Table 1. (continued)

Scientific name	English name	French name	German name	Spanish name	African name	Nr.
Solanum aethiopicum	nightshade, Ethiopian leaves	tomate amère, aubergine indigène	Nachtschatten	morela	Ouolof: diakhatu Bambara: goyo	55, 74
solanum melongea	eggplant, eggfruit, brinjal, aubergine	aubergine	Eierfrucht	berenjena		56
Sorghum spp.	sorghum, Guinea corn, kaffir corn	sorgho	Hirse, Guinea-Korn	sorgo		1, 10
Sphenostylis stenocarpa	yambean, pemo, girigiri	pomme de terre de Mossi, haricot-igname	Yamsbohne	haba de ñame		21
Spondias mombin (S. lutea)	yellow mombin, hogplum, jobo	monbin jaune, prunier mirobolant	Mombinpflaume, Schweinspflaume	jobo	Ouolof: sob Bambara: ninkom Baoulé: troma	106
Tacca involucrata (T. pinnatifida)	tacca, Fiji arrow-root	tacca, arrow-root polynésien	Tahiti Arrow-Root	arrurruz de Polinesia		22
Talinum triangulare	fameflower, waterleaf, Lagos bologi	talinum		espinaca de Filipinas		75
Trianthema postulacacastrum (T. pentandra)	horse-purslane	rotule de chameau "pourpier"	Portulak	verdolaga	Ouolof: oumogelem	76

Table 1. (continued)

Scientific name	English name	French name	German name	Spanish name	African name	Nr.
Triticum spp.	wheat	blé	Weizen	trigo		14
Urena lobata	cadillo, bun-okra, aramina plant	urena	Kongo-Jute	urena		77
Vigna unguiculata spp. catjang	cowpeas, catjang, Hindu cowpeas, kaffir bean, blackeyed pea	niébé, catjang, loubia, haricot dolique	Chinabohne	frijol de ojo negro	Ouolof: niébé Sérère: ognaou Bambara: sosso Sonrhai: doungouni	29, 78
Vitex cuneata	blackplum, West African plum	prune noire		ciruela negra	Ouolof: heul Sérère: dyob Bambara: koto Mossi: andéga	107
Voandzia subterranea	Bambara groundnut, Congo goober, earthpea, kaffir pea, groundbean, jugo bean	pois bambara, voandzon, pois de terre, pois souterrain	Erdnuß	guisante bambara	Ouolof: guerté bambara Bambara: tiga ni nguélé tiga ni n'kourou Baoulé: koro nkoro Lam: souoné	30
Zea mays	maize, corn	maïs	Mais	maíz		3, 13
Ziziphus mauritania	Jujube, India fruit	jujube	Jujube	jujube	Ouolof: dem, dim, siddem Sérère: ngitj Bambara: tomonom Mossi: mougounouga	108

Table 2. Production of cereals, starchy roots and tubers in the French-speaking West Africa (1963/1964)

| Land of production | Cereals | | | Starchy roots and tubers | | | | |
	Nr. 1 Millet and sorghum t	2 Rice paddy t	3 Maize Corn t	4 Manioc Cassava t	5 Yam t	6 Taro t	7 Sweet potato t	8 Plantain t
Ivory Coast	78,000	220,000	178,000	980,000	1,860,000	116,000	56,000	1,068,000
Dahomey	73,000	738	206,000	1,226,000	495,000	—	60,000	—
Upper Volta	1,043,000	30,000	109,000	25,000	29,000	—	42,000	—
Mali	936,000	190,000	70,000	151,000	30,000	—	—	—
Mauritania	60,000	200	3,000	—	—	—	2,000	—
Niger	1,320,000	12,900	2,230	136,000	—	—	24,000	—
Senegal	450,000	89,000	27,000	153,000	—	—	14,700	—
Togo	131,000	17,700	84,000	1,088,000	857,000	750	4,100	—

Table 3. Cereals and grain products
Introduction to tables 3—8

kcal = kilo-calories for 100 g; W = water %; P = protein %; F = fat %; CH = carbohydrate %; B_1 = thiamin mg/100 g; B_2 = riboflavin mg/100 g; C = vitamin C mg/100 g; Ni = niacin mg/100 g; A-equiv. = Vitamin A equivalent in micrograms; Ca = calcium mg/100 g; Fe = iron mg/100 g; Ph = phosphorus mg/100 g.

These tables are only meant to give an approximate orientation. For detailed information, we recommend the following three excellent references:

1. Food Composition Table for Use in Africa (1968), compiled by Woot-Tsuen Wu Leung, US Department of Health, Education and Welfare, Bethesda, Md. With the cooperation of F. Busson and Cl. Jardin, Food Consumption and Planning Branch, Nutrition Division, Food and Agriculture Organisation of the United Nations, Rome.

2. Lexikon der tropischen, subtropischen und mediterranen Nahrungs- und Genußmittel, E.-G. Schenk und G. Naundorf, Manualia Nicolai, Herford, 1966.

3. Aliments de l'Ouest Africain, Tables de composition.
J. Toury, R. Giorgi, J. C. Favier and J. F. Savina O.R.A.N.A. Dakar (1967).

In compiling the following tables, we have amply consulted all three publications. We wish to thank the authors for having permitted us to use their data.

Nr.	Scientific name	English, French, German and Spanish name	kcal	W	P	F	CH	B_1	B_2	C	Ni	A-equiv.	Ca	Fe	Ph
9	Pennisetum spp.	millet, mil, Neger-Hirse, panizo													
		whole grain	340	12	10	4	72	0.3	0.2	—	1.7	—	22	12	290
		milled	265	30	7	2	60	0.2	0.1	—	0.9	—	20	6	260
10	Sorghum spp.	sorghum, sorgho, Guinea-Korn, sorgo													
		whole grain (aver.)	345	10	11	3	74	0.3	0.1	—	3.3	—	26	10	290
		milled	272	29	8	2	60	0.3	0.1	—	2.6	—	13	6	220

Table 3 (continued)

Nr.	Scientific name	English, French, German and Spanish name	kcal	W	P	F	CH	B₁	B₂	C	Ni	A-equiv.	Ca	Fe	Ph
11	Digitaria exilis	fonio, fonio, Findi-Getreide, fonio													
		whole grain	332	10	7	3	74	0.2	0.1	—	2	—	41	9	190
		milled	332	12	6	1	79	0.1	0.1	—	1	—	18	8	80
12	Oryza sativa	rice, riz, Reis, arroz													
		paddy	350	10	6	2	76	0.3	0.05	—	6	—	27	8	323
		polished	368	11	7	1	80	0.2	0.04	—	2	—	14	4	233
13	Zea mays	corn, maïs, Mais, maíz													
		yellow, USA	364	10	10	5	74	0.3	0.1	—	2	—	13	5	220
		white	357	11	9	4	74	0.3	0.1	—	2	—	16	4	220
		sifted	368	12	9	3	74	0.3	0.1	—	1	—	18	3	178
14	Triticum spp.	wheat, blé, Weizen, trigo													
		unsifted	332	12	12	2	73	0.4	0.1	—	4	—	54	6	279
		sifted (70 to 80% extr.)	364	12	11	1	75	—	—	—	—	—	29	4	117

Table 4. *Starchy roots and tubers*

new: Vit. A equivalent in micrograms.   tr. = trace

Nr.	Scientific name	English, French, German and Spanish name	kcal	W	P	F	CH	B₁	B₂	C	Ni	A-equiv.	Ca	Fe	Ph
15	Manihot utilissima	cassava, manioc amer, Maniok, yuca													
		tuber raw	149	62	1.2	0.2	36	0.04	0.05	31	0.6	26	68	2	42
		dried	355	9	2	0.6	85	0.1	0.02	1	1.5	tr.	102	1	98
		meal	344	13	1.6	0.5	83	0.06	0.05	4	1	0	66	4	135
16	Dioscorea alata	yam, igname, Yams, ñame													
		tuber raw	133	64	2	0.1	33	0.05	0.03	15	0.5	6	27	2	56
		flowerflour	335	14	3	0.4	80	0.1	0.1	—	1	—	20	1	110
17	Ipomea batatas	sweet potato, patate douce, Batate, batatas													
		pale	121	69	2	0.2	28	0.1	0.04	37	0.7	75	33	2	38
		yellow	121	69	2	0.2	28	0.1	0.04	37	0.7	1255	33	2	38
18	Colocasia esculenta	taro, taro, Taro, ñampí	105	73	2	0.1	24	0.1	0.02	7	0.8	tr.	83	2	97
19	Musa paradisiaca	plantain, banane plantain, Mehlbanane, plátano	135	65	1	0.3	32	0.1	0.04	20	0.6	780	8	1	38

Table 4 (continued)

Nr.	Scientific name	English, French, German and Spanish name	kcal	W	P	F	CH	B₁	B₂	C	Ni	A-equiv.	Ca	Fe	Ph
20	Solanum tuberosum	potato, pomme de terre, Kartoffel, patata	82	78	2	0.1	19	0.1	0.03	21	1	25	10	1	51
21	Sphenostylis stenocarpa	yambean, pomme de terre du Mossi, Yamsbohne, haba de ñame	136	64	4	0.1	31	0.2	—	20	—	—	8	—	93
22	Tacca involucrata	Fiji arrow-root, arrow-root polynésien, Tahiti arrow-root, arrurruz de Polinesia	121	69	1.5	0.1	29	—	—	13	—	—	8	—	65
23	Cyperus esculentus	tiger-nut, souchet comestible, Cyperngras, chufa	329	31	3	13	52	0.3	0.1	3	1	—	28	7	202

Percentage of losses during processing:

15 Cassava	17%
16 Yam	16%
17 Sweet potato	17%
19 Plantain	32%
20 Potato	14%
all others, about	16%

Table 5. *Grain legumes*

Nr.	Scientific name	English, French, German and Spanish name	kcal	W	P	F	CH	B₁	B₂	C	Ni	A-equiv.	Ca	Fe	Ph
24	Arachis hypogea	peanut, arachides, Erdnuß, cacahuete													
		raw grains	580	4	25	50	18	1	0.1	0	15	12	42	7	350
		roasted grains	592	2	26	51	18	0.4	0.1	0	15	—	42	—	350
		press-cake, village	393	11	37	17	31	1	0.2	0	24	—	80	20	560
		press-cake, factory	364	5	46	6	39	2	—	0	25	—	83	17	580
25	Cajanus cajan	pigeon pea, ambrevade, Taubenerbse, guandú	352	8	20	1	67	1	0.1	0	4	—	150	15	290
26	Dolicos lablab	lablab beans, lablab, Lablab Bohne, lablab	344	10	23	1	63	0.6	0.1	0	2	—	75	—	290
27	Kerstingiella geocarpa	ground bean, lentille de terre, —, lenteja	348	9	20	1	67	0.8	0.2	0	2	—	90	15	390
28	Phaseolus acutifolius	Tepary-bean, haricot-tépary, Tepary-Bohne, judía	353	9	20	1	68	0.3	0.1	0	3	—	110	—	310

Table 5 (continued)

Nr.	Scientific name	English, French, German and Spanish name	kcal	W	P	F	CH	B_1	B_2	C	Ni	A-equiv.	Ca	Fe	Ph
29	Vigna unguiculata	cow peas, niébé, Chinabohne, frijol de ojo negro	346	10	21	1	63	1	0.2	0	3	35	90	9	370
		—Niébé fritters	404	31	8	27	31	0.2	0.05	0	1	—	50	3	150
30	Voandzia subterranea	Bambara groundnut, pois bambara, Erdnuß, guisante bambara	369	10	21	6	60	0.6	0.1	0	2	—	50	16	315

Table 6. *Nuts and seeds*

Nr.	Scientific name	English, French, German and Spanish name	kcal	W	P	F	CH	B_1	B_2	C	Ni	A-equiv.	Ca	Fe	Ph
31	Adansonia digitata	baobab, baobab, Affenbrotbaum, baobab raw grains	445	7	35	29	19	1	0.1	0	1.4	tr.	238	12	1540
32	Anacardium occidentale	cashew, noix d'acajou, Elephantenlaus, marañón roasted nut	554	5	22	44	27	0.5	0.3	0	1.6	tr.	53	11	550
33	Balanites aegyptiaca	desert date, datte du désert, Wüsten-Dattel, dátil del desierto	554	4	27	45	21	2	0.1	0	1.3	0	140	7	525
34	Boscia senegalensis	aisen, boscia, Boscia, boscia raw grains	338	11	24	2	60	0.03	0.03	5	8.8	80	132	10	330
35	Chrysobalamus orbicularis	cocoplum, mafuli, Ikakopflaume, hicaco fruit seeds	449	23	5	23	38	0.5	—	0	—	190	136	4	140

Table 6 (continued)

Nr.	Scientific name	English, French, German and Spanish name	kcal	W	P	F	CH	B₁	B₂	C	Ni	A- equiv.	Ca	Fe	Ph
36	Citrullus vulgaris (lanatus)	watermelon, pastèque, Wassermelone, sandia roasted seeds	578	4	32	50	11	0.05	0.1	0	2	—	44	13	700
37	Cocos nucifera	coconut, noix de coco, Kokosnuß, coco mature kernel	466	41	3.5	40	15	0.03	0.03	2	0.6	12	9	3	135
38	Cola acuminata	colanut, noix de cola, Kolanuß, nuez de cola	186	52	4	0.1	42	0.05	0.02	14	0.75	—	30	2	80
39	Parinari macrophylla	Cayor apple, pomme du Cayor, —, manzana de cayor kernel, dried	650	3	18	64	13	0.5	0.1	4	0.5	105	80	6	530
40	Parkia biglobosa	African locust bean, arbre à farine, Heuschreckenbohne, algarroba africana	432	7	35	22	32	0.5	—	6	—	—	230	11	500
41	Ricinodendron heudolotii	manketti nut, ezezam, Issagnila Nüsse, nuez de ben fermented seeds	431	15	35	29	16	tr.	tr.	0	tr.	0	260	—	480

Table 7. *Vegetables and leaves*

1. Cultivated vegetables

Nr.	Scientific name	English, French, German and Spanish name	kcal	W	P	F	CH	B₁	B₂	C	Ni	A-equiv.	Ca	Fe	Ph
42	Allium cepa	onion, oignon, Zwiebel, cebolla	31	91	1	0.1	7	0.02	0.04	8	0.2	0	23	2	45
43	Brassica oleracea	cabbage, chou pommé, Kohl, colinabo	32	90	1	0.1	8	0.05	0.04	55	0.3	51	53	3	76
44	Brassica oleracea botrytis	cauliflower, chou-fleur, Blumenkohl, coliflor	26	91	2	0.1	6	1	0.1	66	0.5	7	20	2	70
45	Capsicum frutescens	red pepper, piment enragé Cayenne Pfeffer, ají													
		fresh	101	72	4	2	20	0.3	0.1	73	2	4600	72	2	100
		dry	341	10	12	11	62	0.3	0.8	8	6	7000	200	18	330
46	Capsicum annuum	sweet pepper, poivron, Paprika, ají													
		red	40	88	2	0.2	9	0.1	0.2	125	2	1400	38	2	57
		green	32	89	2	0.3	8	0.1	0.1	120	2	85	25	3	60

Table 7 (continued)

Nr.	Scientific name	English, French, German and Spanish name	kcal	W	P	F	CH	B₁	B₂	C	Ni	A-equiv.	Ca	Fe	Ph
47	Cucumis sativus	cucumber, concombre, Gurke, pepino	12	96	1	0.1	3	0.02	0.01	13	0.3	tr.	11	1	38
48	Cucumis melo	musk melon, cantaloup, Zuckermelone, melón tuna leaves	18	94	1	0.1	5	0.04	0.02	11	0.4	15	26	2	36
49	Cucurbita pepo	squash, courge, Kürbis, calabaza	24	93	1	0.1	6	0.04	0.02	6	0.5	1900	27	2	40
50	Daucus carota	carrot, carotte, Karotte, zanahoria	40	89	1	0.1	10	0.03	0.04	6	0.7	3000	27	3	54
51	Hibiscus esculentus	okra, gombo, Gombo, quingombó raw	36	89	2	0.03	9	0.05	0.1	25	0.7	95	70	1	81
		dried, powdered	280	11	11	0.7	69	tr.	0.3	10	4	94	880	34	400

Table 7 (continued)

Nr.	Scientific name	English, French, German and Spanish name	kcal	W	P	F	CH	B_1	B_2	C	Ni	A-equiv.	Ca	Fe	Ph
52	Hibiscus sardaniffa	red sorrel, oseille de Guinée, Rosella, rosa de Jamaica													
		fresh leaves	44	86	2	0.1	11	0.05	0.06	14	0.5	142	162	4	60
		dried	252	18	8	0.3	64	0.03	0.2	10	3	63	1140	31	195
53	Lycopersicum esculentum	tomato, tomate, Tomate, tomate													
		raw	23	93	1	0.2	5	0.1	0.04	31	0.5	180	15	2	23
		cherry	21	93	1	0.1	5	0.05	0.04	48	0.7	520	34	2	60
54	Phaselus spp.	bean, haricot vert, Bohne, judia	26	91	2	0.1	6	0.1	0.1	17	0.5	375	32	3	50
55	Solanum aethiopicum	nightshade, tomate amère, Nachtschatten, morela	30	90	7	0.1	7	0.1	0.1	9	0.6	176	28	1	47
56	Solanum melongena	eggplant, aubergine, Eierfrucht, berenjena	25	92	1	0.2	6	0.1	0.1	5	0.5	17	17	3	21

Table 7 (continued)

2. Leaves

Nr.	Scientific name	English, French, German and Spanish name	kcal	W	P	F	CH	B_1	B_2	C	Ni	A-equiv.	Ca	Fe	Ph
57	Adansonia digitata	baobab, baobab, Affenbrotbaum, baobab leaves dried	279	12	12	3	3	0.1	0.1	tr.	4	4850	2270	25	260
58	Amaranthus hibridus	amaranth, amaranthe, Amaranth, bledo	42	84	5	0.2	8	0.05	0.4	64	1.2	5700	410	9	100
59a	Amaranthus lividus	amaranth, amaranthe, Amaranth, bledo	39	85	4	0.1	8	0.05	0.4	84	1	2400	440	9	95
59b	Amaranthus spinosus	amaranth, amaranthe, Amaranth, bledo	51	81	6	0.2	10	—	—	52	—	4000	430	10	130
60	Arachis hypogea	peanut, arachide, Erdnuß, cacahuete	73	77	5	0.4	16	0.2	0.6	110	2	4000	250	4	90
61	Bauhinia reticulata	gemsbock bean, bauhinia, Bauhinia, bauhinia	64	78	5	0.1	14	—	—	68	—	—	435	—	80

Table 7 (continued)

Nr.	Scientific name	English, French, German and Spanish name	kcal	W	P	F	CH	B₁	B₂	C	Ni	A-equiv.	Ca	Fe	Ph
62	Cassia tora	senna sickle, casse foetide, Fetid-Gemüse, casia	60	80	6	0.1	13	0.2	0.5	120	1.5	3700	610	6	95
63	Ceratotheca sesamoides	false-sesame, ceratotheca, Karkashi-Samen, ceratoteca	50	80	4	0.2	11	—	—	28	—	—	175	—	75
64	Corchorus olitorius	jute, corête potagère, indischer Flachs, corete	67	77	5	0.3	15	0.2	0.5	100	1.5	3000	380	8	135
65	Crataeva adansonii	crateva, crateva, —, crateva	74	76	6	0.1	15	0.4	0.7	200	1.7	2500	420	8	175
66	Cucurbita pepo	squash, courge, Kürbis, calabaza, leaves	24	88	6	tr.	3	—	—	82	—	—	550	—	150
67	Ficus gnaphalocarpa	fig, figue, Feige, higo, leaves	57	80	6	0.2	11	—	—	35	—	—	75	16	160

Table 7 (continued)

Nr.	Scientific name	English, French, German and Spanish name	kcal	W	P	F	CH	B_1	B_2	C	Ni	A-equiv.	Ca	Fe	Ph
68	Gynandropsis pentaphylla	spiderherb, gynandropsis, —, ginandropsis	31	87	6	0.3	4	—	—	13	—	—	285	6	130
69	Hibiscus sabadariffa	red sorrel, oseille de Guinée, Rosella, rosa de Jamaica	47	85	3	0.2	10	0.1	0.3	50	1.2	1950	215	5	90
70	Ipomaea batatas	sweet potato, patate douce, Batate, batatas	47	83	5	0.1	10	0.1	0.3	70	1	3500	170	4	90
71	Leptadenia lancifolia	leptadenia, leptadenia, —, leptadenia	55	80	5	0.1	12	0.2	0.3	75	2	2400	400	5	100
72	Manihot utilissima	cassava, manioc amer, Maniok, yuca leaves	96	70	8	0.6	14	0.2	0.7	370	2	6000	380	8	130

Table 7 (continued)

Nr.	Scientific name	English, French, German and Spanish name	kcal	W	P	F	CH	B₁	B₂	C	Ni	A.-equiv.	Ca	Fe	Ph
73	Moringa pterygosperma	horseradish tree, ben ailé, Meerrettichbaum, ben alado	50	75	8	0.6	14	0.2	0.8	220	3	5500	530	12	140
74	Solanum aethiopicum	nightshade, tomate amère, Nachtschatten, morela	56	81	5	0.4	11	0.2	0.4	67	1.8	3200	500	6	100
75	Talinum triangulare	fameflower, talinum, —, espinaca de Filipinas	47	85	4	0.1	10	0.1	0.2	43	0.3	—	140	4	80
76	Trianthema portulacastrum	horse purslane, rotule de chameau, Portulak, verdolaga	27	90	3	0.2	5	0.05	0.1	30	1	3000	200	4	80
77	Urena lobata	cadillo, urena, Kongo-Yute, urena	54	82	3	0.2	13	—	—	62	—	—	560	—	67
78	Vigna unguiculata	cowpeas, niébé, Chinabohne, frijol de ojo negro	44	85	5	0.3	8	0.2	0.4	60	2.1	3750	300	6	60

Table 8. *Fruits*

Nr.	Scientific name	English, French, German and Spanish name	kcal	W	P	F	CH	B₁	B₂	C	Ni	A-equiv.	Ca	Fe	Ph
79	Adansonia digitata	baobab, pain de singe, Affenbrot, baobab	280	17	2	0.1	76	0.4	0.1	170	2	20	290	7	96
80	Anacardium occidentale	cashew, pomme de cajou, Elephantenlaus, marañón	53	85	1	0.6	12	0.03	0.2	250	0.3	380	12	2	45
81	Ananas comosus	pineapple, ananas, Ananas, ananás	47	87	0.4	0.1	12	0.1	0.03	34	0.1	90	16	0.4	14
82	Annona muricata	soursoup, corossol épineux, Anone, guanábana	60	84	2	0.04	14	0.1	0.1	26	1	tr.	32	2	48
83	Aphania senegalensis	soapberry tree, cerise du Sénégal, —, cereza senegalesa	100	71	2	0.03	26	0.02	—	69	—	160	22	3	96
84	Balanites aegyptiaca	desert date, datte du désert, Wüsten-Dattel, dátil del desierto, dried	268	21	5	0.1	70	0.3	0.1	46	1.7	—	147	4	60

Table 8 (continued)

Nr.	Scientific name	English, French, German and Spanish name	kcal	W	P	F	CH	B_1	B_2	C	Ni	A-equiv.	Ca	Fe	Ph
85	Borassus aethiopicum	African fan palm, rônier, Palmyra-Palme, palmera de azúcar	43	93	1	0.1	5	0.03	0.01	5	0.2	—	10	1	45
86	Carica papaya	papaya, papaye, Papaya, papaya	39	89	1	0.1	10	0.03	0.03	54	0.4	610	42	1	30
87	Chrysobalanus orbicularis	cocoplum, mafuli, Ikakopalme, hicaco	52	85	1	0.1	14	0.02	0.03	19	0.4	65	45	2	42
88	Cola cordifolia	colanut, noix de cola, Kolanuß, nuez de cola	91	74	1	0.04	24	0.01	—	10	0.4	230	55	2	47
89	Cordyla africana	bushmango, cordyla, Wildmango, cordile	69	80	1	0.1	18	0.02	—	74	0.6	126	27	1	135
90	Detarium micro carpum	sweet dattock, dankh, —, detario dulce	310	11	5	0.4	81	0.03	—	32	—	—	82	2	84

Table 8 (continued)

Nr.	Scientific name	English, French, German and Spanish name	kcal	W	P	F	CH	B₁	B₂	C	Ni	A-equiv.	Ca	Fe	Ph
91	Detarium senegalense	Senegal dattock, détar, Talgbaum, detario senegalés	116	67	2	0.4	30	0.1	0.05	1300	0.6	132	27	3	50
92	Dialium guineense	velvet tamarind, tamarinier blanc, Tamarinde, tamarindo fruit pulp	293	17	4	0.1	78	—	—	—	—	—	196	4	77
93	Diospiros mespiliformis	medlar persimmon, kaki de brousse, Kaki-Feige, palo santo	125	64	3	0.04	34	0.01	—	13	0.2	—	42	2	46
94	Ficus gnapholacarpa	fig, figue, Feige, higo	76	78	2	1	17	0.03	0.06	15	0.5	109	250	6	65
95	Ficus iteophylla	fig, figue, Feige, higo	75	77	3	0.2	18	0.04	—	25	—	118	290	10	75
96	Ficus platyphylla	fig, figue, Feige, higo	63	83	2	0.1	16	0.04	0.04	22	0.3	—	205	4	66
97	Landolphia heudelotii	gumvine, landolphia, —, landolfia	44	84	1	0.1	14	0.04	0.03	11	0.8	tr.	15	2	44

Table 8 (continued)

Nr.	Scientific name	English, French, German and Spanish name	kcal	W	P	F	CH	B₁	B₂	C	Ni	A-equiv.	Ca	Fe	Ph
98	Landolphia senegalensis	gumvine, saba, —, landolfia	70	80	1	0.2	18	0.1	0.03	48	0.5	tr.	51	1	28
99	Mangifera indica	common mango, mangue sauvage, Mango, mango													
		ripe	60	83	1	0.1	16	0.02	0.04	40	0.4	920	29	1	30
		unripe, with skin	47	86	1	0.1	12	0.03	0.06	270	0.3	500	57	1	35
100	Musa sapientium	banana, banane, Banane, banana común	61	81	1	0.1	17	0.03	0.04	7	0.6	52	8	2	34
101	Parinari excelsa	parinarium, parinaire, —, parinaria	116	67	1	0.2	31	—	—	31	—	145	37	2	28
102	Parinari macrophylla	Cayor apple, pomme du Cayor, —, manzana de Cayor	141	60	1	0.4	38	0.03	0.1	95	0.7	—	42	2	54
103	Parkia biglobosa	African locust bean, arbre à farine, Heuschreckenbohne, algarroba africana	305	12	3	0.5	81	1	0.7	255	1	1200	125	4	164

Table 8 (continued)

Nr.	Scientific name	English, French, German and Spanish name	kcal	W	P	F	CH	B₁	B₂	C	Ni	A-equiv.	Ca	Fe	Ph
104	Persea gratissima	avocado, avocat, Avocatobirne, aguacate	104	82	1	9	7	0.05	0.2	17	2.7	450	23	2	56
105	Psidium guajava	guava, goyave, Guayave, guayaba	66	81	1	0.1	17	0.06	0.03	152	1.3	165	42	2	42
106	Spondias mombin	yellow mombin, prunier mirobolant, Mombinpflaume, jobo	41	88	1	0.2	10	0.04	0.03	12	1.4	700	24	1	39
107	Vitex cuneata	blackplum, prune noire, —, ciruela negra	90	74	1	0.1	24	0.02	—	6	—	—	37	2	47
108	Ziziphus mauritiana	jujube, jujube, Jujube, jujube													
		fresh	93	71	2	tr.	25	—	—	66	—	—	51	—	20
		dry	286	17	4	0.1	75	0.03	0.02	24	2	0	210	3	56

References

The numbers following the References are page numbers of this book

Aebi, H.: Möglichkeiten zur biochemischen Feststellung des Ernährungszustandes. Klin. Ernährungslehre **13**, 24 (1964)
21, 34, 35
— (2) See Richterich (1963)
30
— (3) See Richterich (1963)
35
— See Eppenberger (1964)
30
— See von Wartburg (1965)
32
— Inborn errors of metabolism. Ann. Rev. Biochem. **36**, 271 (1967). Palo Alto: Annual Review, Inc. Boyer, P. D., Ed.
30, 32
— Richterich, R.: Aktuelles zur Biologie der Enzyme. Helv. med. Acta **30**, 353 (1963)
24, 29
Alleyne, G. A. O.: The effect of severe PCM on the renal function in Jamaican children. Pediatrics **39**, 400 (1967)
119
— See Picou (1965)
51
— See Waterlow (1966)
5
— Young, V. H.: Adrenocortical function in children with severe PCM. Clin. Sci. **33**, 189 (1967)
6, 103, 119
Allison, J. B.: See Waino (1959)
22
— See Munro (1964)
20
Amasuya, A., Narasinga Rao, B. S.: Plasma amino acid pattern in kwashiorkor and marasmus. Amer. J. clin. Nutr. **21**, 723 (1968)
49

Anand, B. K.: Nervous regulation of food intake. Physiol. Rev. **41**, 677 (1961)
9, 100
Andersen, V., Gerhard, W., Clausen, J.: Enzymes of human leucocytes and erythrocytes: lactic acid dehydrogenase isoenzymes and acid phosphatases. Protides of the biological fluids, vol. II, p. 514 (Peeters, H., Ed.). Amsterdam: Elsevier 1964
35
Anderson, H. L., Benevenga, N. J., Harper, A. E.: Associations among food and protein intake, serine dehydratase, and plasma amino acids. Amer. J. Physiol. **214**, 1008 (1968)
74
Antonowicz, J.: See Frenk (1957)
124
Aronoff, M.: See Wainio (1959)
22
Arroyave, G.: Biochemical signs of mild-moderate forms of protein-calorie malnutrition. Mild-moderate forms of protein-calorie malnutrition. Symposium of the Swedish Nutrition Foundation I (Blix, G., Ed.), Bastad, August 29—31, 1962, p. 32—46. Uppsala: Almqvist & Wiksells 1963
48, 50, 53
— Biochemical characteristics of malnourished infants and children. Proc. Western Hemisphere Nutrition Congress: organised by the Council on Foods and Nutrition, American Medical Association, November 8 to 11, 1965, Chicago, Ill. Amer. med. Ass. 1966, p. 30—36
53, 54
— See Castellanos (1961)
103

Arroyave, G.: See Viteri (1966)
52
— Bowering, J.: Plasma free amino acids as an index of protein nutrition. Arch. lat-amer. Nutr. 18, 341 (1968)
49
— Guillermo, W., de Funes, C.: Razón nitrógeno ureico/creatinina como indicador del nivel de ingesta proteica. II. Diferencias en cuanto a urea urinaria y amonio, con y sin diuresis de agua provocada, en grupos de niños con características dietéticas diferentes. Arch. lat-amer. Nutr. 17, 49 (1967)
53
— Jansen, A. A. J., Torrico, M.: Razón nitrógeno ureico/creatinina como indicador del nivel de ingesta proteica. I. Efecto de la ingesta de agua sobre la excreción "basal" de urea y creatinina de niños con estados nutricionales diferentes. Arch. lat-amer. Nutr. 16, 203 (1966)
53, 54
— Wilson, D., de Funes, C., Behar, M.: The free amino acids in blood plasma of children with kwashiorkor and marasmus. Amer. J. clin. Nutr. 11, 517 (1962)
49
— — Viteri, F.: Variations in urine and blood serum nitrogenous constituents with controlled protein intakes. VIIth International Congress of Nutrition, p. 164. Hamburg, 3—10 August, 1966
53
Auricchio, S.: See Shmerling (1964)
116
Ayyoub, N.: See McLaren (1965)
49

Bachhawat, B. K.: See Ittyerah (1967)
36
Banerjee, G.: See Weber (1961)
23
Barbezat, W.: See Saunders (1967)
49
Barnes, D. J.: See Nichols (1968)
119

Behar, M.: See Arroyave (1962)
49
— See Viteri (1966)
52, 113
Beloff-Chain, A., Catanzaro, R., Chain, E. B., Masi, I., Pocchiari, F.: Fate of uniformly labelled C-glucose in brain slices. Proc. roy. Soc. B 144, 22 (1955)
134
Benedict, F. G.: See Wright, S., Applied physiology, p. 810. London: Oxford University Press 1915
8
Bender, A. E.: Private communication during a discussion at the meeting of the Group of European Nutritionists, Cambridge 1968
40
Bendicenti, A., Mariani, A., Paolucci, A. M., Spadoni M. A.: Captazione della L-S metionina da parte del fegato di ratti adulti in rapporto al contenuto in proteine della dieta. Quaderni della nutrizione 19, No. 6, 217 (1959)
5
Benevenga, N. J., Harper, A. E., Rogers, Q. R.: Effect of an amino acid imbalance on the metabolism of the most-limiting amino acid in the rat. J. Nutr. 95, 434 (1968)
74
— See Anderson (1968)
74
Bengoa, J. M.: Priorities in public health nutrition problems. Proceedings of the 7th International Congress of Nutrition, Hamburg, 1966, vol. 4, p. 811. Braunschweig: F. Vieweg & Sohn Verlag 1967
11
Bensusan, C.: Fibre formation from solutions of collagen. — II. The role of tyrosyl residues. J. Amer. chem. Soc. 82, 4990 (1960)
83
Bernstein, E.: See Wainio (1959)
22
Black, E.: See Hoffenberg (1962)
118

Bowes, J. H., Elliot, R. G., Moss, J. A.:
Recent advances in Gelalin and Glue
research, p. 71. Quoted in "Treatise
on collagen" 1, 66 (Ramachandran,
G. N., Ed.). New York: Academic
Press 1958
83

Bowering, J.: See Arroyave (1968)
49

Brand, K.: See Hess (1965)
19

Brock, J. F.: The results of chronic
dietary malnutrition and undernutrition (as seen in the health of
different racial groups in Africa).
Humanity and Subsistence, Symposium, 1960. Vevey: Annales
Nestlé 1961
58

— See Hoffenberg (1962)
118

Bronstein, S. B.: See Weber (1961)
23

Bücher, T.: Enzyme unter biologischem
Aspekt. Erbliche Stoffwechselkrank-
heiten (Linneweh, E., Ed.). München/
Berlin: Urban and Schwarzenberg
1962
24, 25

— See Pette (1963)
24, 30

Burger, A.: See Richterich (1963)
30

Burgess, E. A.: See Solimano (1967)
4

Busson, F.: See Wu Leung (1968)
158

Campbell, V. S.: See Fox (1968)
64

Cannon, W. B.: Physiological regulation
of normal states: some tentative
postulates concerning biological homeostasis. A Charles Richet: ses
amis, ses collègues, ses élèves, 22 mai
1926, p. 91—93. Paris: Ed. Méd.
1926.
1, 6, 9

Carbonara, A. D.: See Mancini (1965)
118

Castellanos, H., Arroyave, G.: Role of
adrenal cortical system in the re-

sponse of children to severe PCM.
Amer. J. clin. Nutr. 9, 186 (1961)
103

Catanzaro, R.: See Beloff-Chain (1955)
134

Cellis', W. R. F.: See Edozien (1960)
49

Chaikoff, I. L.: See Fitch (1960)
23, 24

Chain, E. B.: See Beloff-Chain (1955).
134

Chan, H.: See Waterlow (1966)
5

— Waterlow, J. C.: The protein requirement of infants at the age of
about one year. Bull. Wld Hlth Org.
33, 881 (1965)
11

— — The protein requirement of infants at the age of about one year.
Brit. J. Nutr. 20, 775 (1966)
11

Clausen, J.: See Andersen (1964)
35

Cook, G. C., Kajubi, S. K.: Tribal incidence of lactase deficiency in
Uganda. Lancet 1966 I, 725—729.
42

— Lakin, A., Whitehead, R. G.: Absorption of lactose and its digestion
products in the normal and malnourished Ugandan. Gut. 8, 622 to
627 (1967)
42

Couvée, L. M. J., Nugteren, D. H.,
Luyken, R.: The nutritional condition of the Kapaukus in the Central Highlands of Netherlands New
Guinea. I. Biochemical examina-
tions. Trop. geogr. Med. 14, 27
(1962)
53

Cowley, J. J., Griesel, R. D.: The effect
on growth and behavior of rehabili-
tating first and second generation
low-protein rats. Anim. Behav. 14,
506 (1966)
134

Cravioto, J.: See Frenk (1957)
124

— See Westall (1958)
49

Cravioto, J.: See Ramos-Galván (1958)
129

— Gómez, F., Ramos-Galván, R., Frenk, S., Montaño, E. L., Garcia, N.: Metabolismo proteico en la desnutrición avanzada: Concentración de aminoácidos libres en el plasma sanguíneo. Bol. Ofic. sanit. panamer. 48, 383 (1960)
49

Cravioto, C. J.: See Waterlow (1960)
34, 35

Darby, W. J.: See Sandstead (1965)
5

Dawson, D. M., Eppenberger, H. M., Kaplan, N. O.: Creatine kinase: Evidence for a dimeric structure. Biochem. biophys. Res. Commun. 21, 346 (1965)
30

DeMayer, E. M.: See Vis (1965)
42

— FAO/WHO/UNICEF guidelines for safety evaluation and human testing of supplementary food mixtures, in a monograph on protein-enriched cereal foods for world needs. To be published in 1969 by the American Association of Cereal Chemists
12

Dubach, U. C.: Enzymes in urine and kidney. Current problems in clinical biochemistry. Bern/Stuttgart: Hans Huber Verlag 1968
35, 119

Dubois, R.: See Vis (1965)
42

— Vis, H., Loeb, H., Vincent, M.: Altérations de l'aminoacidurie observées dans des cas de kwashiorkor. Helv. paediat. Acta 14, 13 (1959)
120

Dugdale, A. E., Edkins, E.: Urinary urea/creatinine ratio in healthy and malnourished children. Lancet 1964 I, 1062
111

Dumm, M. E.: See Ittyerah (1965)
49

— See Ittyerah (1967)
36

Durand, P.: Lattosuria idiopatica in una paziente con diarrea cronica ed acidosi. Minerva pediat. 10, 27 (1958)
116

Edkins, E.: See Dugdale (1964)
111

Edozien, J. C., Phillips, E. J., Cellis', W. R. F.: The free amino acids of plasma and urine in kwashiorkor. Lancet 1960 I, 615
49

Eisenberg, quoted by Evang, K.: Health of Mankind, p. 97. Ciba Foundation Symposium. March 1967. London: J. & A. Churchill 1967
61

Elliot, R. G.: See Bowes (1958)
83

Elvehjem, C. A.: See Williams (1949)
22

Eppenberger, M.: See Eppenberger, H. M. (1964)
30

Eppenberger, H. M.: See Dawson (1965)
30

— Eppenberger, M., Richterich, R., Aebi, H.: The ontogeny of creatine kinase isoenzymes. Develop. Biol. 10, 1 (1964)
30

Evang, K.: Discussion on cancer and cardiovascular diseases. Health of Mankind, p. 97. Ciba Foundation Symposium. March 1967. London: J. & A. Churchill 1967
61

FAO Nutritional Studies No. 16. Protein requirements, report of the FAO Committee, Rome 1957
12

FAO. Third World Food Survey. Freedom from hunger campaign. Basic Study No. 11. Rome: FAO, Italy 1963
98

Favier, J. C.: See Toury (1967)
158

Fellingham, S. A.: See Wittman (1967)
15

Fernandez-Varela, H.: See Valenzuela (1966)
124

Fitch, W. M., Chaikoff, I. L.: Extent and patterns of adaptation of enzyme activities in livers of normal rats fed diets high in glucose and fructose. J. biol. Chem. **235**, 554 (1960)
23, 24

Folin, O.: Approximately complete analyses of thirty "normal" urines. Amer. J. Physiol. **13**, 45 (1905)
2

— Laws governing the chemical composition of urine. Amer. J. Physiol. **13**, 66 (1905)
2

— A theory of protein metabolism. Amer. J. Physiol. **13**, 117 (1905)
2

Fox, H. C., Campbell, V. S., Morris, J. C.: The dietary and nutritional status in Jamaican infants and toddlers. Information 8, 33 (1968)
64

Freedland, R. A., Murad, S., Hurvitz A. I.: Relationship of nutritional and hormonal influences on liver enzyme activity. Fed. Proc. **27**, 5, 1217 (1968)
28

Frenk, S.: See Cravioto (1960)
49

— Metcoff, J., Gómez, J., Ramos-Galván, R., Cravioto, J., Antonowicz, J.: Intracellular composition and homeostatic mechanisms in severe infantile malnutrition. — II. Composition of tissues. Pediatrics **20**, 105 (1957)
124

Funes de, C.: See Arroyave (1962)
49

— See Arroyave (1967)
53

Gabr, M. K.: See Sandstead (1965)
5

Gaetani, S., Paolucci, A. M., Spadoni, M. A., Tomassi, G.: Activity of amino acid activating enzymes in tissues from protein depleted rats. J. Nutr. **84**, 173 (1964)
5

Gaitonde, M. K., Richter, D.: Protein synthesis in the rat brain. J. Biochem. **55**, 8 (1953)
134

Garcia, N.: See Cravioto (1960)
49

Garcia Antillon, L.: See Gordillo (1957)
119

Garfinkel, D., Hess, B.: Metabolic control mechanisms. VII. A detailed computer model of the glycolytic pathway in ascites cells. J. biol. Chem. **239**, 971 (1964).
24

Garrow, J. S.: (2) See Waterlow (1966)
5

Gerhard, J. C., Pardee, A. B.: The enzymology of control by feed-back inhibition. J. biol. Chem. **237**, 891 (1962)
20

Gerhard, W.: See Andersen (1964)
35

Giorgi, R.: See Toury (1967)
158

Gómez, F.: See Frenk (1957)
124

— See Westall (1958)
49

— See Cravioto (1960)
49

Gopalan, C.: See Vasantha (to be published)
80

Gordillo, G., Soto, R., Metcoff, J., Lopez, E., Garcia Antillon, L.: Intracellular composition and homeostatic mechanisms in severe chronic infantile malnutrition. III. Renal adjustments. Pediatrics **20**, 303 (1957)
119

Gordon, J. E.: See Mata (1967)
16

— See Scrimshaw (1968)
16

Griesel, R. D.: See Cowley (1966)
134

Guillermo, W.: See Arroyave (1967)
53

Hadorn, B.: See Shmerling (1964)
116
Hansen, J. D. L.: See Pimstone (1966)
103
— See Truswell (1966)
45, 49
— See Saunders (1967)
49
— See Wittman (1967)
15
Harper, A. E.: Balance and imbalance of amino acids. Ann. N. Y. Acad. Sci. **69**, 1025—1038 (1958)
70
— Effect of variations in protein intake on enzymes of amino acid metabolism. Canad. J. Biochem. **43**, 1589—1603 (1965)
74
— See Yoshida (1966)
74
— See Benevenga (1968)
74
— See Anderson (1968)
74
— Diet and plasma amino acids. Amer. J. clin. Nutr. **21**, 358—366 (1968)
74
— Rogers, Q. R.: Amino acid imbalance. Proc. Nutr. Soc. **24**, 173 to 190 (1965)
73
— Leung, P., Yoshida, A., Rogers, Q. R.: Amino acid balance and imbalance: XI. Some new thoughts on amino acid imbalance. Fed. Proc. **23**, 1087—1092 (1964)
73
Harris, H.: Enzyme polymorphisms in man. Proc. roy. Soc. B **164**, 298 (1966)
30
— Enzyme variations in man: some general aspects. Proceedings of the 3rd International Congress of Human Genetics, p. 207, (Crow, J. F., Neel, J. V., Eds.). Baltimore: The John Hopkins Press 1967
30, 32
— Hopkinson, D. A., Luffman, J. E., Rapley, S.: Electrophoretic variation in erythrocyte enzymes. Hereditary disorders of erythrocyte metabolism, p. 1, (Beutler, E., Ed.). New York/London: Grune and Stratton 1968
30, 31
Hay, A.: (2) See Waterlow (1966)
5
Hazlewood, C. F.: See Nichols (1968)
119
Heard, C. R. C., Kriegsman, S. M., Platt, B. S.: The interpretation of plasma amino acid ratios in protein-calorie deficiency. Proc. Nutr. Soc. **27**, 20A (1968)
40
Heremans, J. F.: See Mancini (1965)
118
Hernández-Peniche, J.: See Valenzuela (1966)
124
Hess, B.: See Garfinkel (1964)
24
— Brand, K.: Enzyme action in living cells. Clin. Chem. **11**, 223 (1965)
19
Hifney El, A.: See Sandstead (1965)
5
Hodge, A. J.: Quoted in International Review of Connective Tissue Research, vol. I, p. 64, (Hall, D., Ed.) (1960)
83
Hoffenberg, R., Saunders, S., Linder, G. C., Black, E., Brock, J. F.: Protein metabolism, an International Symposium, p. 314 (Gross, F., Ed.). Berlin-Göttingen-Heidelberg: Springer 1962
118
Holt, L. E.: See Westall (1958)
49
— Protein economy in the growing child. Postgrad. Med. **27**, 783 (1960)
60
Hopkinson, D. A.: See Harris (1968)
30, 31
Howells, G. R., Wharton, B. A., McCance, R. A.: Value of hydroxyproline indices in malnutrition. Lancet **1967** I, 1082
51
— See Wharton (1967)
51

Ittyerah, T. R., Pereira, S. M., Dumm, M. E.: Serum amino acids of children on high and low protein intakes. Amer. J. clin. Nutr. 17, 11 (1965)
49
— Dumm, M. E., Bachhawat, B. K.: Urinary excretion of lysosomal arysulfatases in kwashiorkor. Clin. chim. Acta 17, 405 (1967)
36

James, P.: See Waterlow (1966)
5
Jansen, A. A. J.: See Arroyave (1966)
53, 54
Jardin, Cl.: See Wu Leung (1968)
158
Jelliffe, D. B., Symonds, B. E. R., Jelliffe, E. F. P.: The pattern of malnutrition in early childhood in Southern Trinidad. J. Pediat. 57, 922 (1960)
64
Jelliffe, E. F. P.: See Jelliffe, D. B. (1960)
64
Jones, J. D.: See Neldner (1966)
81

Kajubi, S. K.: See Cook (1966)
42
Kanul, W. W.: See McLaren (1965)
49
Kaplan, N. O.: See Dawson (1965)
30
Katsunuma, H.: Occupational health and its assessment. Health of Mankind, p. 113. Ciba Foundation Symposium. March 1967. London: J. & A. Churchil, 1967
61
Kerpel-Fronius, E., Varga, F., Kun, K., Vonoczky, J.: The relationship between circulation and kidney function in infantile dehydration and malnutrition. Acta med. Acad. Sci. hung, 5, 27 (1954)
119
Kidley, P. T.: See Nasset (1967)
100

Kosterlitz, H. W.: Effects of dietary protein on liver cytoplasma. Nature (Lond.) 154, 207 (1944)
22
Krauss, M., Mayer, J.: Influence of protein and amino acids on food intake in the rat. Amer. J. Physiol. 209, 479 (1965).
101
Kremzner, L. T.: See Wainio (1959)
22
Kriegsman, S. M.: See Heard (1968)
40
Kun, K.: See Kerpel-Fronius (1954)
119

Lambo, T. A.: Mental and behaviour disorders. Health of Mankind, p. 103. Ciba Foundation Symposium. March 1967. London: J. & A. Churchill 1967
62
Lakin, A.: See Cook (1967)
42
Lát, J.: The spontaneous exploratory reaction. Pharmacology of conditioning, learning and retention. Oxford: M. Mikelson (Ed.) 1965
134
— Widdowson, E. M., McCance, R. A.: Some effects of accelerating growth. III. Behavior and nervous activity. Proc. roy. Soc. B 153, 347 (1960)
134
Lepkovsky, S.: Nutritional stress factors and food processing. Advanc. Food Res. 4, 105 (1953)
135
— Potential pathways in nutritional progress. Food Technol. 13, 421 (1959)
135
Leung, P. M. B.: See Harper (1964)
73
— See Yoshida (1966)
74
Levin, B.: See Solimano (1967)
4
Linder, G. C.: See Hoffenberg (1962)
118
Loeb, H.: See Dubois (1959)
120
Lopez, E.: See Gordillo (1957)
119

Lovell, H. G.: See McKenzie (1967)
64

Luffman, J. E.: See Harris (1968)
30, 31

Luyken, R.: See Couvée (1962)
53

— Luyken-Koning, F. W. M.: Studies on the physiology of nutrition in Surinam. III. Urea excretion. Trop. geogr. Med. **12**, 237 (1960)
53

Luyken-Koning, F. W. M.: See Luyken, R. (1960)
53

Mancini, G., Carbonara, A. O., Heremans, J. F.: Immunochemical quantitation of antigens by single radial immunodiffusion. Immunochem. **2**, 235 (1965)
118

Mariani, A.: See Bendicenti (1959)
5

— Spadoni, M. A., Tomassi, G.: Effect of protein depletion on amino acid activating enzymes of rat liver. Nature (Lond.) **199**, 378 (1963)
5

Markert, C. L., Møller, F.: Multiple forms of enzymes: Tissue, ontogenetic and species specific patterns. Proc. nat. Acad. Sci. (Wash.) **45**, 753 (1959)
29

Masi, I.: See Beloff-Chain (1955)
134

Mata, L. J., Urrutia, J. J., Gordon, J. E.: Diarrhoeal disease in a cohort of Guatemalan village children observed from birth to age two years Trop. geogr. Med. **19**, 247 (1967)
16

Mayer, J.: See Krauss (1965)
101

McCance, R. A.: See Lát (1960)
134

— See Howells (1967)
51

— See Wharton (1967)
51

McKenzie, H. I., Lowell, H. G., Standard, K. L., Miall, W. E.: Child mortality in Jamaica. Milbank mem. Fd. Quart. **45**, 303 (1967)
64

McLaren, D. S., Kanul, W. W., Ayyoub, N.: Plasma amino acids and the detection of PCM. Amer. J. clin. Nutr. **17**, 152 (1965)
49

Mendes, C. B., Waterlow, J. C.: The effect of a low-protein diet, and of refeeding, on the composition of liver and muscle in the weanling rat. Brit. J. Nutr. **12**, 74 (1958)
6

Metcoff, J.: Ionic composition and cell metabolism with PCM in man. Amer. J. clin. Nutr. **21**, 376 (1968)
34, 35

— See Gordillo (1957)
119

— See Frenk (1957)
124

Miall, W. E.: See McKenzie (1967)
64

Miller, L. L.: Changes in rat liver enzyme activity with acute inanition. J. biol. Chem. **172**, 113 (1948)
22

Mokhtar, N.: See Sandstead (1965)
5

Møller, F.: See Markert (1959)
29

Montaño, E. L.: See Cravioto (1960)
49

Moodie, A. D.: See Wittman (1967)
15

Morris, J. C.: See Fox (1968)
64

Moser, H.: See Wiesmann (1965)
35

Moss, J. A.: See Bowes (1958)
83

Munro, H. N.: Mammalian protein metabolism, vol. 1, chap. 10 (Munro, H. N., Allison, J. B., Eds.). New York/London: Academic Press, Inc. 1964
6, 20

— Role of amino acid supply in regulating ribosome function. Fed. Proc. **27**, 5, 1231 (1968)
20

Munro, H. N., Allison, J. B.: Nutritional aspects of protein metabolism. Patho logical aspects of protein metabolism, vol. 2 (Munro, H. N., Allison, J. B., Eds.). New York/London: Academic Press, Inc. 1964
20

Murray, P.: See Pimstone (1966)
103

Narasinga Rao, B. S.: See Amasuya (1968)
49

Nasset, E. S., Kidley, P. T., Schenk, E. A.: Hypothalamic lesions related to ingestion of an imbalanced amino acid diet. Amer. J. Physiol. **213**, 645 (1967)
100

Naundorf, G.: See Schenk (1966)
158

Neldner, K. H., Jones, J. D., Winkelmann, R. K.: Scleroderma dermal amino acid composition with particular reference to hydroxyproline. Proc. Soc. exp. Biol. (N. Y.) **122**, 39 (1966)
81

Nichols, B. L., Hazlewood, C. F., Barnes, D. J.: Percutaneous needle biopsy of quadriceps muscle. Potassium analysis in normal children. J. Pediat. **72**, 840 (1968)
119

Nirenberg, M.: Protein synthesis and the RNA code. Harvey Lect. 59. New York: Academic Press 1965
4

Nugteren, D. H.: See Couvée (1962)
53

Norton, P. M.: Plasma amino acid levels in kwashiorkor., (abstract) p. 7. Proceedings of the 5th International Congress of Nutrition. Washington, D. C., September 1st to 7th, 1960
49

Olson, R. E.: Are we looking at the right enzyme system ? Amer. clin. Nutr. **20**, 604 (1967)
4, 35

Papenberg, J.: See von Wartburg (1965)
32

Paolucci, A. M.: See Bendicenti (1959)
5

— See Gaetani (1964)
5

Pardee, A. B.: See Gerhard (1962)
20

Peña de la, C.: See Westall (1958)
49

Pereira, S. M.: See Ittyerah (1965)
49

Pette, D., Bücher, T.: Proportionskonstante Gruppen in Beziehung zur Differenzierung der Enzymaktivitätsmuster von Skelet-Muskeln des Kaninchens. Hoppe-Seyler's Z. physiol. Chem. **331**, 180 (1963)
24, 30

Phillips, E. J.: See Edozien (1960)
49

Picou, D.: (2) See Waterlow (1966)
5

— Alleyne, G. A. O., Seakins, A.: Hydroxyproline and creatinine excretion in infantile protein malnutrition. Clin. Sci. **29**, 517 (1965)
51

Pimstone, B. L., Wittman, W., Hansen, J. D. L., Murray, P.: Growth hormone and kwashiorkor. — Role of protein in growth hormone homeostasis. Lancet **1966**, 779
103

Platt, B. S.: Malnutrition in African mothers, infants and young children. Report of the 2nd International African Conference on Nutrition, p. 153. Gambia, November 19th to 27th, 1952. London: Her Majesty's Stationery Office 1954
53

— See Heard (1968)
40

Pocchiari, F.: See Beloff-Chain (1955)
134

Potter, V. R.: Systematic oscillation in metabolic functions in liver from rats adapted to controlled feeding schedules. Fed. Proc. **27**, 5, 1238 (1968)
29

Prader, A.: See Shmerling (1964)
116
Prasad, A. S.: See Sandstead (1965)
5

Ramos-Galván, R.: See Frenk (1957)
124
— See Cravioto (1960)
49
— Cravioto, J.: Desnutrición en el niño. Concepto y ensayo de sistematización. Bol. méd. Hosp. infant. (Méx.) **20**, 169 (1958)
129
Rao, K. J.: Plasma cortisol levels in protein-calorie malnutrition. Arch. Dis. Childh. **43**, 365 (1968)
84
Rapley, S.: See Harris (1968)
30, 31
Rasmussen, H.: See Westall (1958)
49
Rein, H.: Einführung in die Physiologie des Menschen. Berlin-Göttingen-Heidelberg: Springer 1948
3
Richter, C. P.: Total self-regulatory functions in animals and human beings. Harvey Lect. **38**, 63 (1942).
133
Richter, D.: The turnover of proteins and lipids in vivo in the brain. "Neurochemistry". Springfield: Elliott (Ed.) 1962
134
— See Gaitonde (1953)
134
Richterich, R.: See Aebi (1963)
24, 29
— Burger, A.: (1) Lactic dehydrogenase isoenzymes in human cancer cells and malignant effusions. Enzym. biol. clin. **3**, 65 (1963)
30
— Schafroth, P., Aebi, H.: (2) A study of lactic dehydrogenase isoenzyme pattern of human tissues by absorption-elution on Sephadex-DEAE. Clin. chim. Acta 8, 178 (1963)
30
— Rosin, S., Aebi, H., Rossi, E.: (3) Progressive muscular dystrophy. V.

The identification of the carrier state in the Duchenne type by serum creatine kinase determination. Amer. J. hum. Genet. **15**, 133 (1963)
35
— See Eppenberger (1964)
30
— See Wiesmann (1965)
35
Rogers, Q. R.: See Harper (1964)
73
— See Harper (1965)
73
— See Yoshida (1966)
74
— See Benevenga (1968)
74
Roitman, E.: See Westall (1958)
49
Rosin, S.: (3) See Richterich (1963)
35
Rossi, E.: (3) See Richterich (1963)
35
— See Wiesmann (1965)
35
Rubino, A.: See Shmerling (1964)
116
Rutishauser, I. H. E., Whitehead, R. G.: Field evaluation of two biochemical tests which may reflect nutritional status in three areas of Uganda. Brit. J. Nutr. **23**, 1 (1969)
49

Safwat Shukry, A.: See Sandstead (1965)
5
Sandstead, H. H., Safwat Shukry, A., Prasad, A. S., Gabr, M. K., El Hifney, A., Mokhtar, N., Darby W. J.: Kwashiorkor in Egypt. I. Clinical and biochemical studies with special reference to plasma zinc and serum lactic dehydrogenase. Amer. J. clin. Nutr. **17**, 15 (1965)
5
Saunders, S.: See Hoffenberg (1962)
118
Saunders, S. J., Truswell, A. S., Barbezat, G. O., Wittman, W., Hansen, J. D. L.: Plasma free amino acid pattern in protein-calorie malnutri-

tion; reappraisal of its diagnostic
value. Lancet **1967** II, 795
49

Savina, J. F.: See Toury (1967)
158

Schafroth, P.: (2) See Richterich (1963)
30

Schenk, E. A.: See Nasset (1967)
100

Schenk, E.-G., Naundorf, G.: Lexikon
der tropischen, subtropischen und
mediterranen Nahrungs- und Genußmittel. (Manualia Nicolai, Ed.).
Herdorf 1966
158

Schimke, R. T.: (1) Adaptive characteristics of urea cycle enzymes in the
rat. J. biol. Chem. **237**, 459 (1962)
5

— (2) Differential effects of fasting and
protein-free diet on levels of urea
cycle enzymes in rat liver. J. biol.
Chem. **237**, 1921 (1962)
23, 25

— Protein turnover and the regulation
of enzyme levels in rat liver. International Symposium on Enzymatic
Aspects of Metabolic Regulation.
Mexico-City 1966
25, 27

— Regulation of protein turnover in
mammalian tissues. Fed. Proc. **27**,
5, 1223 (1968)
26

Schürch, P. M.: See von Wartburg (1968)
32, 33

Scrimshaw, N. S., Taylor, C. E., Gordon,
J. E.: Interactions of nutrition and
infection. WHO monograph series
No. 57 (1968)
16

Seakins, A.: See Picou (1965)
51

Shmerling, D. H., Auricchio, S., Rubino, A., Hadorn, B., Prader A.: Der
sekundäre Mangel an intestinaler
Disaccharidaseaktivität bei der
Cöliakie. Helv. paediat. Acta **19**, 507
(1964)
116

Solimano, G., Burgess, E. A., Levin, B.:
Protein-calorie malnutrition: effect

of deficient diets on enzyme levels
of jejunal mucosa of rats. Brit. J.
Nutr. **21**, 55 (1967)
4

Soto, R.: See Gordillo (1957)
119

Spadoni, M. A.: See Bendicenti (1959)
5

— See Mariani (1963)
5

— See Gaetani (1964)
5

Srikantia, S. G.: See Vasantha (to be
published)
80

Standard, K. L.: See McKenzie (1967)
64

Stephen, J. M. L.: See Waterlow (1960)
34, 35

— (1) See Waterlow (1966)
5

— (2) See Waterlow (1966)
5

— Adaptive changes in liver and muscle
of rats during protein depletion and
refeeding. Brit. J. Nutr. **22**, 153
(1968)
5

— Waterlow, J. C.: Effect of malnutrition on activity of two enzymes
concerned with amino acid metabolism in human liver. Lancet **1968**
I, No. 7534, 118
5, 34

Symonds, B. E. R.: See Jelliffe, D. B.
(1960)
64

Taylor, C. E.: See Scrimshaw (1968)
16

Tomassi, G.: See Mariani (1963)
5

— See Gaetani (1964)
5

Torrico, M.: See Arroyave (1966)
53, 54

Toury, J., Giorgi, R., Favier, J. C.,
Savina, J. F.: Aliments de l'Ouest
africain. Tables de composition.
O.R.A.N.A., Dakar (1967). See also
Ann. Nutr. Aliment. **26**, 73 (1967)
158

Truswell, A. S.: See Saunders (1967)
49
— Wannenburg, P., Wittman, W.,
Hansen, J. D. L.: Plasma amino
acids in kwashioɪkor, Lancet **1966 I**,
1162
45, 49

Urrutia, J. J.: See Mata (1967)
16

Valenzuela, R. H., Hernández-Peniche,
J., Fernández-Varela, H.: Umbral
electroencefalográfico de excitabi-
lidad y de terminación de serotonina
en niños sanos y desnutridos de
tercer grado. Memoirs, VIII Pan-
american, I Latinamerican and XI
Mexican Congress of Pediatɪics,
p. 346. Mexico-City 1966
124

Vanderborght, H.: See Vis (1965)
42

Varga, F.: See Kerpel-Fronius (1954)
119

Vasantha, L., Skrikantia, S. G., Gopalan, C.: Biochemical changes in skin
in kwashioɪkor. Amer. J. clin. Nutr.
(to be published)
80

Vincent, M.: See Dubois (1959)
120

Vis, H. L.: See Dubois (1959)
120
— Dubois, R., Vanderborght, H.,
DeMayer, E.: Etude des troubles
électrolytiques accompagnant le
kwashiorkor marastique. Rev. franç.
Etud. clin. biol. **10**, 279 (1965)
42

Viteri, F. E.: See Arroyave (1966)
53
— (unpublished data)
52
— Arroyave, G., Behar, M.: Estimation of protein depletion in malnourished children by a creatinine-
height index. VIIth International
Congress of Nutrition, p. 46. Hamburg: August 3—10th, 1966
52, 113

Vonoczky, J.: See Kerpel-Fronius (1954)
119

Waddington, C. H.: The strategy of genes
London: Allen & Unwin (Eds.) 1957
110

Wainio, W. W., Allison, J. B., Kremzner, L. T., Bernstein, E., Aronoff,
M.: Enzymes in protein depletion.
III. Enzymes of brain, kidney,
skeletal muscle, and spleen. J. Nutr.
67, 197 (1959)
22

Wannenburg, P.: See Truswell (1966)
45, 49

von Wartburg, J. P., Schürch, P. M.:
Atypical human liver alcohol dehydrogenase. Conference on Pharma-
cogenetics, The New York Academy
of Sciences. Ann. N. Y. Acad. Sci.
151, 936 (1968)
32, 33
— Papenberg, J., Aebi, H.: An atypical
human alcohol dehydrogenase.
Canad. J. Biochem. **43**, 889 (1965)
32

Waterlow, J. C.: Fatty liver disease in
infants in the British West Indies.
Spec. Rep. Med. Res. Comm. (London). No. 263 (1948)
64
— See Mendes (1958)
6
— Effect of protein depletion on the
distribution of protein synthesis.
Nature (Lond.) **184**, 1875 (1959)
5
— Protein nutrition and enzymes changes in man. Fed. Proc. **18**, 1143
(1959)
5
— See Stephen (1968)
5, 34
— See Chan (1965, 1966)
11
— Stephen, J. M. L.: (1) Adaptation
of the rat to a low-protein diet: the
effect of a reduced protein intake on
the pattern of incorporation of ^{14}C-L-
lysine. Brit. J. Nutr. **20**, 461—484
(1966)
5

Waterlow, J. C., Cravioto, C. J., Stephen, J. M. L.: Protein malnutrition in man. Advanc. Protein Chem. 15, 131 (1960)
34, 35
— Alleyne, G. A. O., Chan, H. V., Garrow, J. S., Hay, A., James, P., Picou, D., Stephen, J. M. L.: (2) Observations on the mechanisms of adaptation to the low protein intakes. Arch. lat.-amer. Nutr. 16, 175 (1966)
5
Weber, R., Banerjee, G., Bronstein, S. B.: Role of enzymes in homeostasis. III. Selective induction of increases of liver enzymes involved in carbohydrate metabolism. J. biol. Chem. 236, 3106 (1961)
23
Westall, R. G., Roitman, E., de la Peña, C., Rasmussen, H., Cravioto, J., Gómez, F., Holt, L. E.: The plasma amino acids in malnutrition. Preliminary observations. Arch. Dis. Childh. 33, 499 (1958)
49
Wharton, B. A.: See Howells (1967)
51
— Howells, G. R., McCance, R. A.: Hydroxyproline indices. Nature (Lond.) 215, 968 (1967)
51
Whipple, C. H.: Hemoglobin, plasma protein and cell protein. Springfield, Ill.: Thomas 1948
5
Whitehead, R. G.: Rapid determination of some plasma amino acids in sub-clinical kwashiorkor. Lancet 1964 I, 250
49, 112
— Amino acid metabolism in kwashiorkor. I. Metabolism of histidine and imidazole derivatives. Clin. Sci. 26, 271 (1964)
40
— Hydroxyproline-creatinine ratio as an index of nutritional status and rate of growth. Lancet 1965 II, 567
51, 52, 112
— Biochemical tests in differential diagnosis of protein and calories

deficiencies. Arch. Dis. Childh. 42, 479 (1967)
44
— See Cook (1967)
42
— See Rutishauser (1969)
49
White House, The: Report of the President's Science Advisory Committee. The World Food Problem, Vol. I, p. 26, B 2 (1967)
7
WHO: Malnutrition and disease. Freedom from hunger campaign. Basic Study No. 12. Geneva: WHO (Switzerland) (1963)
97
— Tech. Rep. Ser. 301. Protein requirements Report of a joint FAO/WHO Expert Group, Geneva 1965
12
— Tech. Rep. Ser. 377. 7th Report of a joint FAO/WHO Expert Committee on Nutrition, Geneva 1967
16
Widdowson, E. M.: See Lát (1960)
134
Wiesmann, U., Moser, H., Richterich, R., Rossi, E.: Progressive Muskeldystrophie. VII. Die Erfassung von Heterozygoten der Duchenne-Muskeldystrophie durch Messung der Serum Kreatin-Kinase unter lokalisierter Arbeitsbelastung in Anoxie. Klin. Wschr. 43, 1015 (1965)
35
Williams, D. C.: Kwashiorkor. Nutritional disease of childhood associated with maize diet. Lancet 1935 II, 1151
80
Williams, J. N., Elvehjem, C. A.: The relationship of amino acid availability in dietary protein to liver enzyme activity. J. biol. Chem. 181, 559 (1949)
22
Wilson, D.: See Arroyave (1962)
49
— See Arroyave (1966)
53
Winkelmann, R. K.: See Neldner (1966)
81

Wittman, W.: See Pimstone (1966)
103
— See Truswell (1966)
45, 49
— See Saunders (1967)
49
— Moodie, A. D., Fellingham, S. A., Hansen, J. D. L.: An evaluation of the relationship between nutritional status and infection by means of a field study. S. Afr. med. J. 41, 664 (1967)
15

Wright, N.: Faulty nutrition: Failures in food supply, variety and distribution. Health of Mankind, Ciba Foundation Symposium, March 1967, p. 151. London: J. & A. Churchill, Ltd. 1967
59

Wu Leung, W.-T., Busson, F., Jardin, Cl.: Food Composition Tables. Publication of US Department of Health, Education and Welfare, Bethesda, Md. Food Consumption and Planning Branch, Nutrition Division, FAO, Rome (1968)
158

Yoshida, A.: See Harper (1964)
73
— Leung, P. M.-B., Harper, A. E., Rogers, Q. R.: Effect of amino acid imbalance on the fate of the limiting amino acid J. Nutr. 89, 80—90 (1966)
74

Young, V. H.: See Alleyne (1967)
6, 103, 119

Index